Springer-Verlag   D-6900 Heidelberg 1 · Postfach 1780
Telefon (0 62 21) 4 91 01 · Telex 04-61723
D-1000 Berlin 33 · Heidelberger Platz 3
Telefon (03 11) 82 20 01 · Telex 01-83319

Springer-Verlag   New York, NY 10010 · 175, Fifth Avenue
New York Inc.   Telefon 673-2660

38 Fortschritte der chemischen Forschung
Topics in Current Chemistry

Phosphorus-Carbon Double Bonds

Springer-Verlag
Berlin Heidelberg GmbH 1973

ISBN 978-3-540-06164-9   ISBN 978-3-540-38372-7 (eBook)
DOI 10.1007/978-3-540-38372-7

Originally published by Springer-Verlag Berlin Heidelberg New York in 1973

Library of Congress Catalog Card Number 51–5497.

Contents

Delocalized Phosphorus-Carbon Double Bonds

Phosphamethin-cyanines, λ^3-Phosphorins and λ^5-Phosphorins

Prof. Dr. Karl Dimroth

Organische Chemie im Fachbereich Chemie der Universität Marburg/Lahn

Contents

Contents

I. Introduction

Before 1964 no stable compound with a localized or delocalized carbon-phosphorus double bond was known. Indeed, it was generally assumed that the atomic radius of phosphorus, being larger than that of carbon or nitrogen, would not provide sufficient $2_p\pi-3_p\pi$ overlap for such a π system to be stable [1]. Our first communication, written jointly with Peter Hoffmann [2a], which described the synthesis of a stable *"phosphamethin-cyanine"* *1* with a delocalized P–C double bond was therefore received with skepticism [1]. However, after Allmann confirmed the structure by X-ray analysis [3], the existence of a new type of phosphorus bond in a *cationic* delocalized π system was unambigously established [2b].

This development encouraged further attempts to synthesize a *neutral* system with a delocalized P–C double bond, which — by analogy to pyridine — could be termed *phosphorin*[a] or, more systematically, *phosphabenzene*. G. Märkl (1966) [5] first succeeded in preparing 2.4.6-tri-phenylphosphabenzene *2*, or as we shall call it, *2.4.6-triphenyl-λ^3-phosphorin*.

Märkl [6] had previously synthesized a different type of cyclic, unsaturated phosphorus compound *3* $(R^2 = R^4 = R^6 = H; R^1 = R^{1'} = C_6H_5)$, which appeared to have P–C bonding much like the P–C double bond in the phosphonium ylids discovered by Wittig [7]. It was not at first clear whether this compound had an ylid P–C bonding system in which the negative charge is delocalized over the five sp^2–C atoms or whether it represented a new type of fully delocalized 6π-electron aromatic system. Several properties of the unsubstituted 1,1-diphenyl-λ^5-phosphorin, such as high reactivity and basicity, seemed to stress the ylid character. On the other hand, its close relationship to the λ^3-phosphorins as well as the extreme stability of a large series of compounds having hetero atoms at the phosphorus were more in line with a 6π-electron aromatic system. Indeed, many of these compounds fail to display typical ylid reactions. Current arguments support the "aromatic" nature of λ^5-phosphorins *3* as we shall demonstrate.

A comparison of the λ^5-phosphorins with the well known cyclic phosphacenes [8] reveals relatively large differences in their physical and chemical properties.

---

[a] Taken from the Ring Index [4], which suggested the term phosphorin, prior to its discovery. In order to distinguish between *2* and the P,P-substituted phosphorins *3*, we designate compounds of type *2* as λ^3-phosphorins and compounds of type *3* as λ^5-phosphorins. Nomenclature proposed by IU PAC-Commission, see Angew. Chem. *84*, 526 (1972); Angew. Chem. Int. Edit. *6*, 506 (1972).

Introduction

Phosphamethin-cyanine   λ³-Phosphorin   λ⁵-Phosphorin

1   *2*   *3*

Y = S, NR, CH = CH; R^2, R^4, R^6 = H, R^2, R^4, R^6 = H,
X = ClO$_4$, BF$_4$ Alkyl, Aryl   Alkyl, Aryl
                                                                $R^1, R^{1'}$ = Alkyl, Aryl,
                                                                OR, NR$_2$, SR, F

This progress report summarizes the extensive work on the syntheses and properties of compounds of the three classes *1*, *2* and *3*.

II. Phosphamethin-cyanines

A. Synthesis

All presently known phosphamethin-cyanines were prepared according to our original procedure (1964) in which two quaternary salts of a heterocyclic base (e. g. 4) are condensed with tris-hydroxymethyl-phosphine 5 in the presence of a proton-abstracting base [2a, b)]. The preparation of bis-[N-ethyl-benzothiazole(2)]-phosphamethin-cyanine-tetrafluoroborate 6 illustrates the synthetic sequence. A mixture of 2 moles of N-ethyl-2-chlorobenzothiazolium-tetrafluoroborate 4 and 1 mole of tris-hydroxymethyl-phosphine [b)] 5 in dimethylformamide is slowly reacted with ethyl-di-isopropylamine or pyridine at 0 °C. Addition of water immediately affords the crystalline cyanine dye 6 in ca. 45% yield:

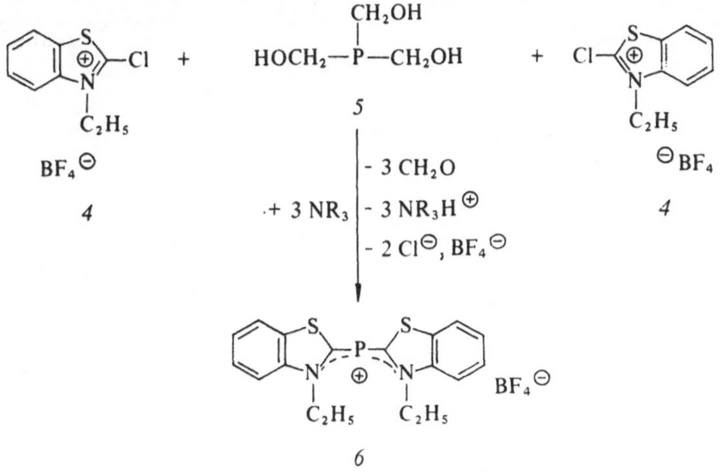

Quaternary salts of other heterocyclic bases can also be employed. For example, symmetrically substituted phosphamethin-cyanines having quinoline and benzimidazole moieties, 7 and 8, respectively, have been prepared [2b, 9)].

[b)] Due to danger of explosion, tris-hydroxymethyl-phosphine should not be distilled. In case of an aqueous solution containing benzene, the water and organic solvent can be removed by vacuum distillation (maximum temperature 60 °C). We thank Farbwerke Hoechst A. G. for generous gifts of aqueous $P(CH_2OH)_3$-solutions.

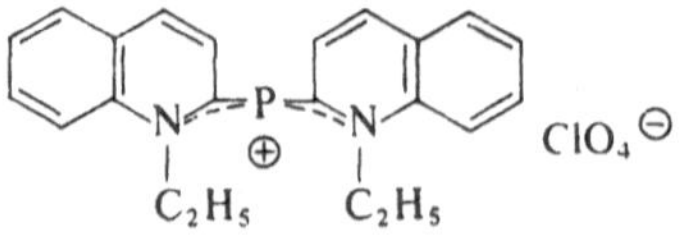

7

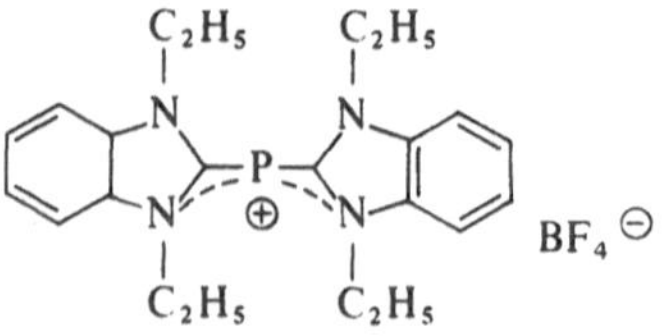

8

Unsymmetrically substituted phospha-methin-cyanines *9* with two different heterocyclic bases can also be synthesized. Here the quaternary salt of one heterocyclic base (*e. g. 4*) is reacted with tris-hydroxymethyl-phosphine *5* in dimethylformamide or glacial acetic acid without the addition of a base. Then one mole of the quaternary salt of the other heterocyclic base (*e. g. 10*) and the base are added. The base deprotonates the hydroxymethyl groups of the phosphine *5*, thus liberating formaldehyde and yielding the phosphine base which then reacts with the second quaternary salt:

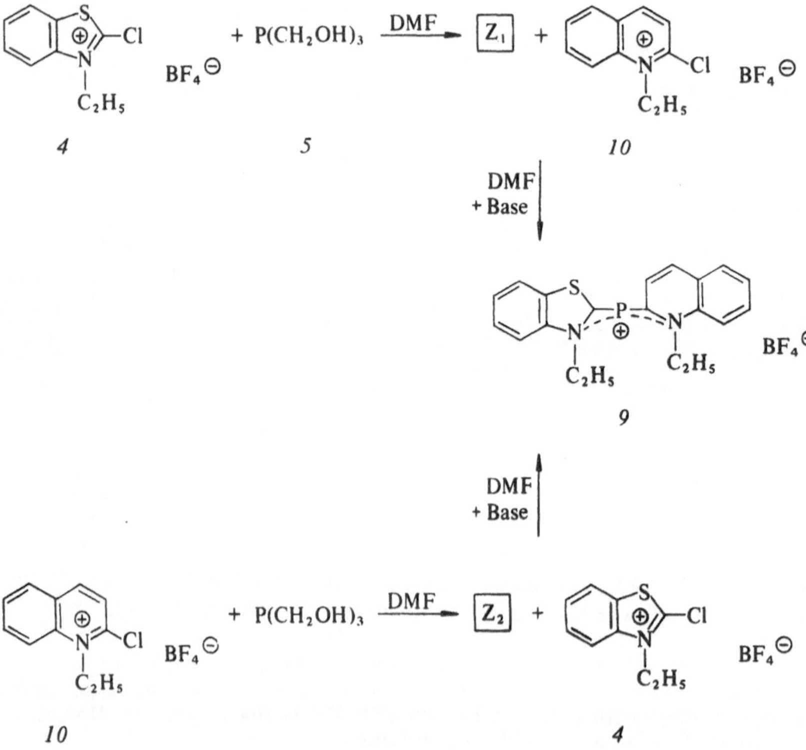

Our assumption that the first step of the reaction involves the formation of an intermediate was confirmed by *Greif* [9], who isolated two such compounds, *12* and *16*, starting from 1,3-dimethyl-2-chloro-benzimidazolium-tetrafluoroborate *11* and 1-methyl-2-chloro-quinolinium-tetrafluoroborate *15*:

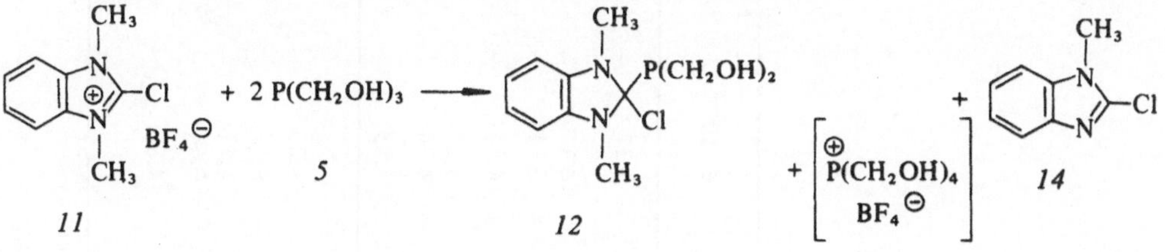

During formation of the addition compounds *12* and *16*, no free formaldehyde accumulates. We assume that the liberated formaldehyde immediately reacts with tris-hydroxymethyl-phosphine, forming the quaternary tetrakis-hydroxymethyl-phosphonium salt *13*. The addition compounds *12* and *16* are relatively unstable, but can be purified for analysis. Intermediates *12* and *16* can also be employed in the synthesis of symmetrical or unsymmetrical phosphamethin-cyanines. For example, Klapproth [11] synthesized *18* in 60% yield by condensing *16* with *11*.

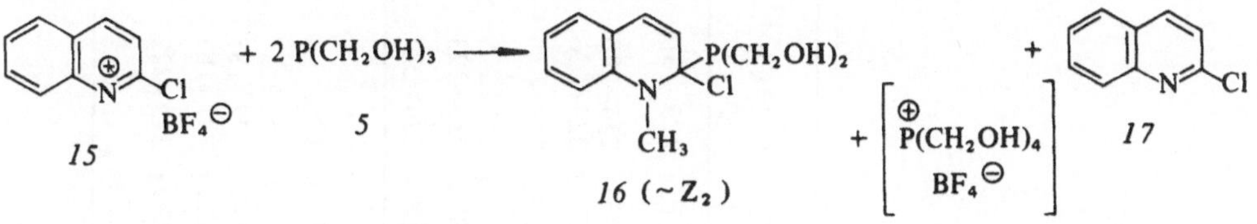

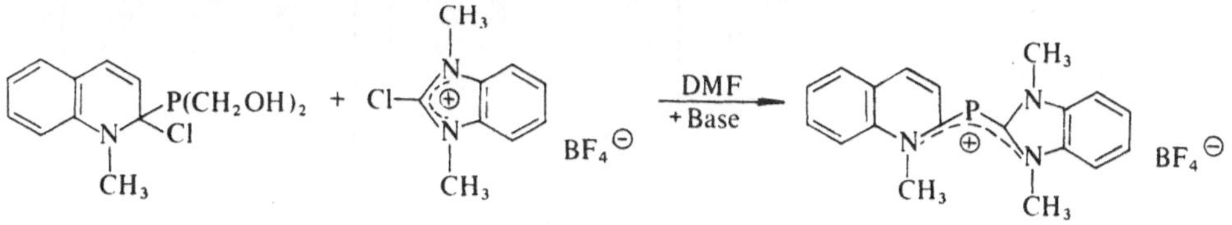

It is likely that intramolecular dealkylation of *12* and *16* leads to *14* and *17*, respectively, since these are sometimes formed in considerable amounts as side-products in the synthesis of phosphamethin-cyanines.

A similar phosphamethin-cyanine synthesis starting from tris-trimethylsilyl-phosphine has been described by Märkl. Using this procedure, arsamethin-cyanines can also be prepared from tris-trimethylsilyl-arsin [10].

Table 1 lists all phosphamethin-cyanines which have been obtained in analytically pure form, together with some of their physical properties.

Table 1. Phosphamethincyanines

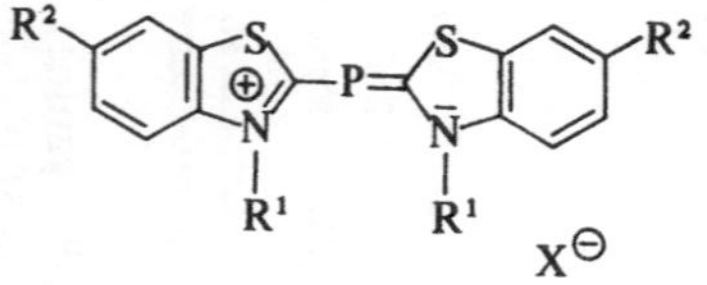

No.	R^1	R^2	R^3	R^4	x	m. p. (°C)	$\lambda_{max.}$ nm	$\epsilon \cdot 10^{-4}$	Solvent	Lit.
6a	CH_3	H	–	–	BF_4	225– 30	472	4.37	a)	2)
6b	C_2H_5	H	–	–	BF_4 ClO_4	214–220 224–226	472 478	4.37 5.14	a) a)	2) 2)
6c	CH_3	OCH_3	–	–	BF_4	220	485	4.34	a)	2)
6d	C_2H_5	OCH_3	–	–	ClO_4 BF_4	213–215 215–223	494 485	3.88 4.34	a) a)	2) 2)
6e	C_2H_5	Br	–	–	ClO_4	224–227	489	4.83	a)	2)
7a	C_2H_5	H	H	H	BF_4 ClO_4	126 197–200	592 605	4.27 5.15	a) a)	2) 2)
7b	C_2H_5	CH_3	H	H	BF_4 ClO_4	178–185 187–189	587 595	4.11 3.64	a) a)	2) 2)
7c	C_2H_5	H	Br	H	ClO_4	219–222	618	4.86	b)	9)

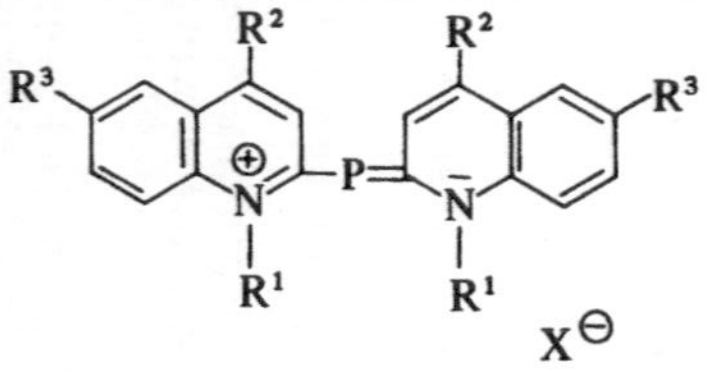

Table 1 (continued)

No.	R¹	R²	R³	R⁴	X	m. p. (°C)	$\lambda_{max.}$ nm	$\epsilon \cdot 10^{-4}$	Solvent	Lit.
8a	CH₃	CH₃	—	—	BF₄	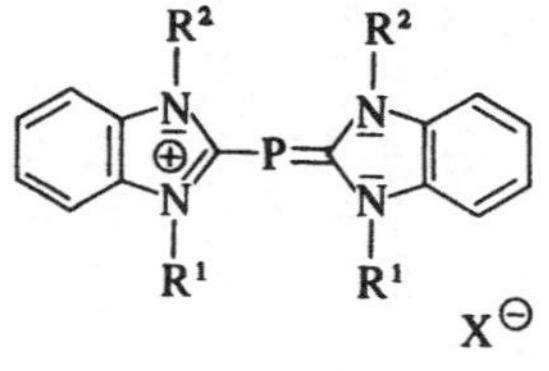 210–211	421 335	3.26	1.14 b)	9)
8b	CH₃	C₂H₅	—	—	BF₄	187–188	424 336	3.24	1.15 b)	9)
8c	CH₃	CH(CH₃)₂	—	—	BF₄	180–181.5	424 347	3.02	1.1 b)	9)
8d	C₂H₅	C₂H₅	—	—	BF₄	162–164	437 344	3.0	1.0 b)	9)
8e	C₂H₅	CH(CH₃)₂	—	—	BF₄	172–173	440 347	2.8	0.94 b)	9)
9a	C₂H₅	H	H	—	ClO₄	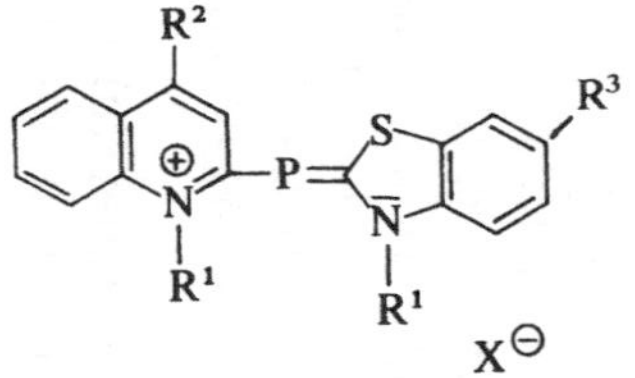 163–167	562 —	3.45	—	a)	2)
9b	C₂H₅	CH₃	H	—	ClO₄	204–210	557 —	3.51	—	a)	2)
9c	C₂H₅	H	OCH₃	—	ClO₄	162–168	565 —	3.60	—	a)	2)
18	CH₃	CH₃	CH₃	—	BF₄	183–190	459 —	2.27	—	c)	11)

a) In 1.2-dichloretane.
b) In dichlormethane.
c) In methanol.

11

B. Physical Properties

The spectral features in the UV and visible regions of phosphamethin-cyanines resemble those of the correspondingly substituted methin- and azamethin-cyanines. The position and extinction coefficient of the maxima as well as the general shape are quite similar (Fig. 1).

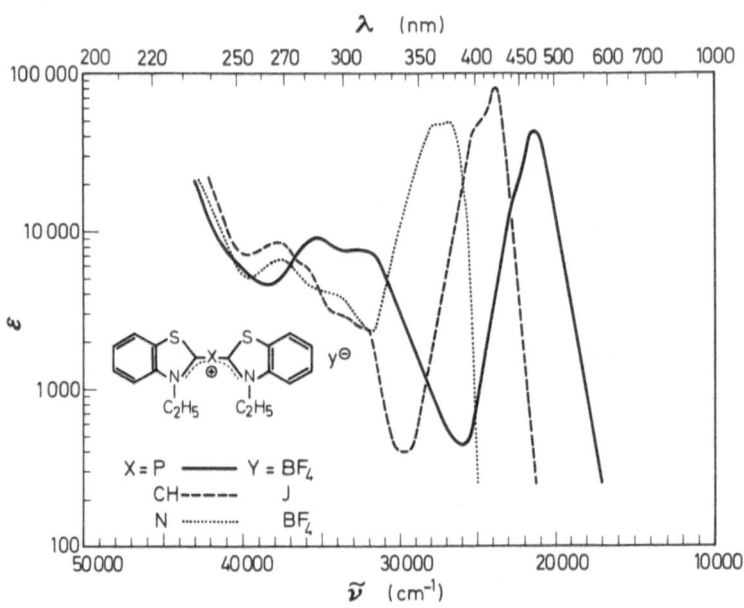

Fig. 1. UV spectra of bis-[N-ethyl-benzthiazole(2)]-phospha-methin-cyanine-tetrafluoro-borate compared with the aza- and the carbomethine-cyanine

In accord with the electron-gas model of cyanines of H. Kuhn [12], phospha-methin-cyanines with the less negative phosphorus absorb at longer wavelengths than the cyanines, in contrast to azamethincyanines with the more negative nitrogen which absorb at shorter wavelengths [13]. The absorption maxima of un-symmetrically substituted phosphamethin-cyanines lie between those of the corresponding symmetric compounds; thus, rules similar to those which Brooker [15] has proposed for methin-cyanines appear to be valid here also. Preliminary studies have also shown that phosphamethin-cyanines can be utilized as sensitizing agents in photography [14].

In the phosphamethin-cyanine series with benzimidazolium substituents, the effect of the size of the N-alkyl groups on the absorption spectra (Fig. 2) was also investigated [9]. Table 1, *8a–8e*, shows that with increasing alkyl size, the long-

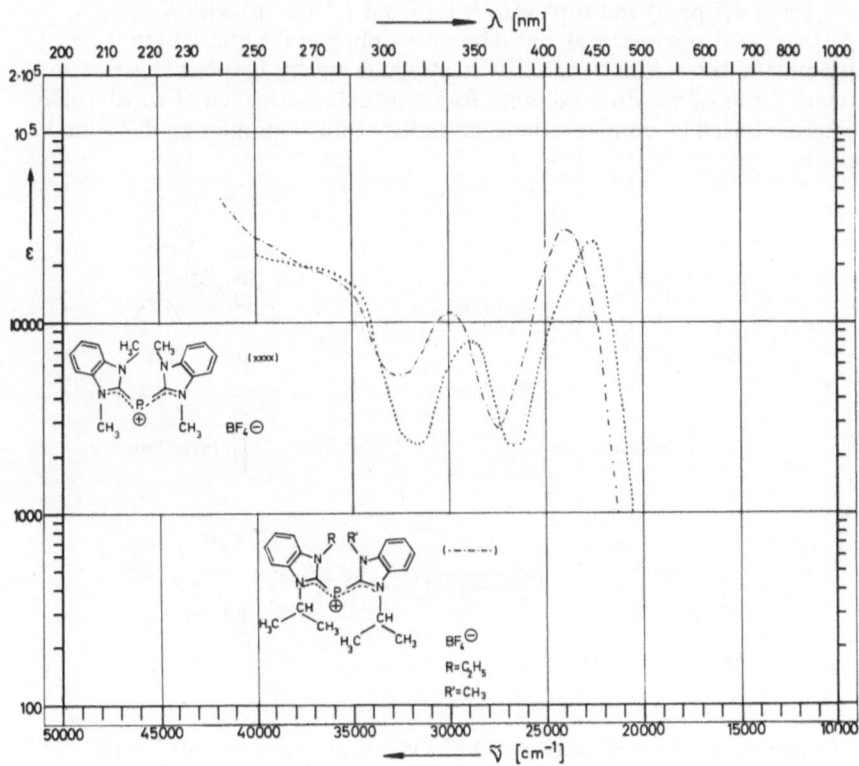

Fig. 2. UV spectra of two different bis-N-[alkylimidazole(2)]-phospha-methin-cyanine-tetrafluoroborates (in dichloromethane)

wave maximum shifts to a longer wavelength with concomitant decrease in extinction. This *"Brunings-Corwin effect"*, well known in cyanine chemistry [16], is explained by the fact that the alkyl groups cause a more pronounced distortion from planarity so that the *ground-state energy* is raised, thereby lowering the excitation energy.

The IR-spectra of phosphamethin-cyanines display no peculiarities [2, 9].

The NMR spectra are characterized by a number of noteworthy features. The methyl protons of bis-[N-methyl-benzthiazole]-phosphamethin-cyanine *6a* in DMF appear as a weakly split signal at $\delta = 4.22$ ppm (TMS standard[c]). The small splitting of 2 Hz could be due to phosphorus coupling. In going to formic acid or trifluoroacetic acid, solvents in which this cyanine is not yet protonated, the ^{1}H-NMR spectrum does not change significantly. In contrast, the spectrum of the ethylcyanine *6b* in formic acid shows two distinctly separated quartets

---

[c] In this review the negative signs of ^{1}Hδ-values have been omitted.

13

(δ = 4,9 and 4,7 ppm) and triplets (δ = 1,15 and 1,55 ppm) with $J_{H-C-C-H}$ = 12,5 Hz, as well as some weak phosphorus coupling in the quartet (ca. 1 Hz). This points to the existence of a sufficiently high energy barrier between two different forms of *6b*. Four different forms are conceivable, all of which could be interconverted by configurational *inversion* at the P-atom or by P–C bond *rotation*:

At present we do not know which of these processes accounts for the NMR spectrum.

In contrast, *bis-[1,3-dimethyl-benzimidazole-2]-phosphamethin-cyanine 8a* shows a single, sharp signal for all four methyl groups (in pyridine at δ = 3,62 ppm) (Fig. 3). In formic acid it shifts to δ = 4,25 ppm, in trifluoroacetic acid to δ = 4,35 ppm. In these acidic media this strongly basic phosphamethin-cyanine is protonated, forming the dication. Due to the steric hindrance between alkyl groups, neither the neutral nor the protonated cyanine dye can be expected to be planar. The situation here is apparently quite similar to that of the corresponding bis-[1,3-dimethyl-benzimidazole-2]carbomethincyanine, which also shows only one signal for the four methyl protons (DMSO solvent) at 3,54 ppm. In the corresponding dication the signal shifts to 4,02 ppm[11].

The two N-methyl groups of 1.2.3-trimethyl-benzimidazoliumiodide in DMSO also absorb at 4,05 ppm, while the signal due to the methyl groups at C_2 appears at 2,9 ppm [11].

In the 1,3-diethyl derivative of the benzimidazolephosphamethin-cyanine series *8d*, one finds the four ethyl groups to be equivalent (pyridine as solvent). The quartet due the CH_2 group ($J_{H-C-C-H}$ = 8Hz) appears at δ = 4,35 ppm, considerably lower than the singlet due to the CH_3 groups in *8a*. The small broadening seems to be due to P-coupling (ca. 1 Hz). The triplet of the CH_3 group lies at δ = 1,25 ppm. In formic acid the quartet and triplet appear at δ = 4,72

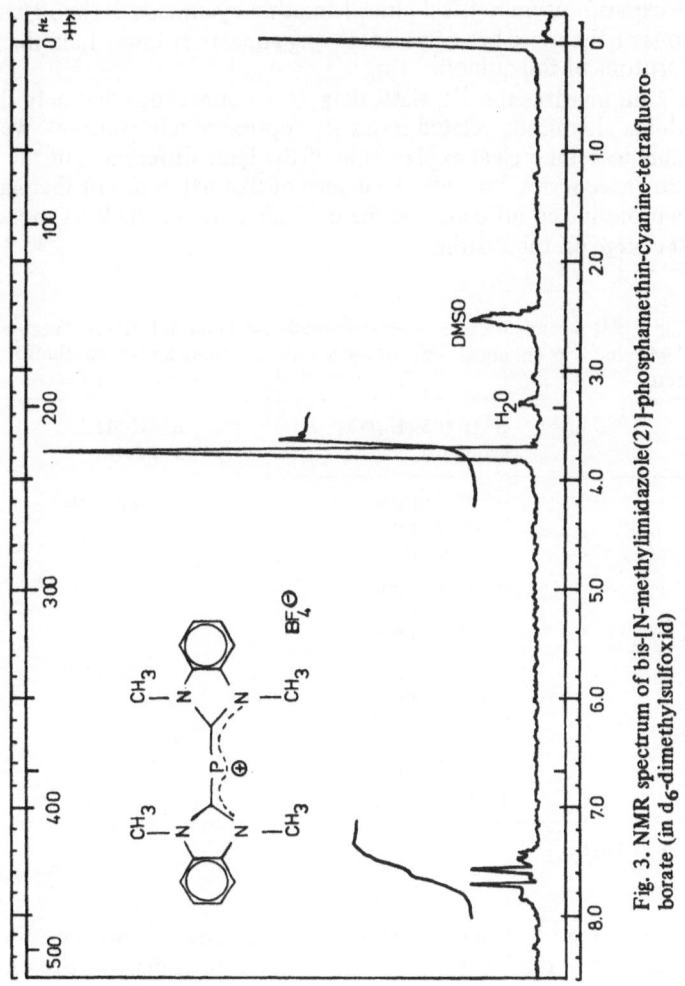

Fig. 3. NMR spectrum of bis-[N-methylimidazole(2)]-phosphamethin-cyanine-tetrafluoro-borate (in d$_6$-dimethylsulfoxid)

and 1,55 ppm, respectively, while the $J_{H-C-C-H}$ coupling remains the same. Phosphamethin-cyanines with different alkyl groups at the two N atoms of both benzimidazole rings show two distinct signals (in formic acid). For example, in formic acid the CH$_2$ protons of the ethyl groups in *8e* appear as a quartet at δ = 4,62 ppm and the CH protons of the isopropyl groups as a multiplet at ca. δ = 5,3 ppm. The signals of the CH$_3$ groups belonging to the isopropyl substituent appear at δ = 1,75 ppm (6H) and the CH$_3$ groups of the ethyl substituents at δ = 1,45 ppm (3H).

No temperature dependence has yet been observed. However, the solubility of the phosphamethin-cyanines is generally so limited that little variation is possible with respect to solvent, concentration or temperature. Comparative experiments with corresponding methin- and azamethin-cyanines are currently under way in our laboratories[11].

15

In the case of unsymmetrical phosphamethin-cyanines *18*, the signal of the CH_3 protons of the more basic imidazole ring appears at lower field than that of the CH_3 protons of the quinoline ring.

Table 2 summarizes the ^{31}P-NMR data. It is noteworthy that only the δ-values of ^{31}P within a chemically related series are approximately constant. We have not yet been able to offer a clear explanation of the large differences of the ^{31}P-shifts between the three series. Possibly the degree of distortion out of the plane of the rings has a pronounced influence on the delocalization of the P = C bond and on the partial charge on the P-atom.

Table 2. ^{31}P chemical shifts of phosphamethin-cyanine tetrafluoroborates in DMF and ^{13}C chemical shifts of analogous methin-cyanine tetrafluoroborates

	δ^{31}P (85%H$_3$PO$_4$ standard)	δ^{13}C (TMS standard)
6a	− 26,05 ppm	− 82,6 ppm
6b	− 24,9 ppm	
6c	− 22,0 ppm	
7a	− 57,1 ppm	− 92,3 ppm [a)]
7b	− 48,8 ppm	
7c	− 66,6 ppm	
8a	+ 103,8 ppm	− 49,0 ppm
8b	+ 104,9 ppm	
8c	+ 97,9 pmm	
8d	+ 112,1 ppm	
8e	+ 109,2 ppm	
18	+ 16,3 ppm	

[a)] N-methyl.

By investigating ^{13}C labeled methin-cyanine analogs [11)], we have found that the chemical shift of the ^{13}C signals show a quite similar dependence on the nature of the basic ring [d)].

C. Results of X-Ray Analysis

Allmann [3)] and Kawada and Allmann [17)] carried out X-ray analysis on bis-[N-ethyl-benzthiazole]-phosphamethin-cyanine perchlorate *6b* and bis-[N-ethyl-quinoline]-phosphamethin-cyanine perchlorate *7b*. In *6b* the two heterocycles lie roughly in the same plane. The C−P−C bond angle is 104,6°. The P−C bond lengths, 1,75$_4$ and 1,75$_7$ Å, can be regarded as identical, proving that the C−P−C bond network is

[d)] We have to thank Prof. Dr. W. v. Philipsborn, University of Zürich, for the ^{13}C−NMR-data and helpful discussion.

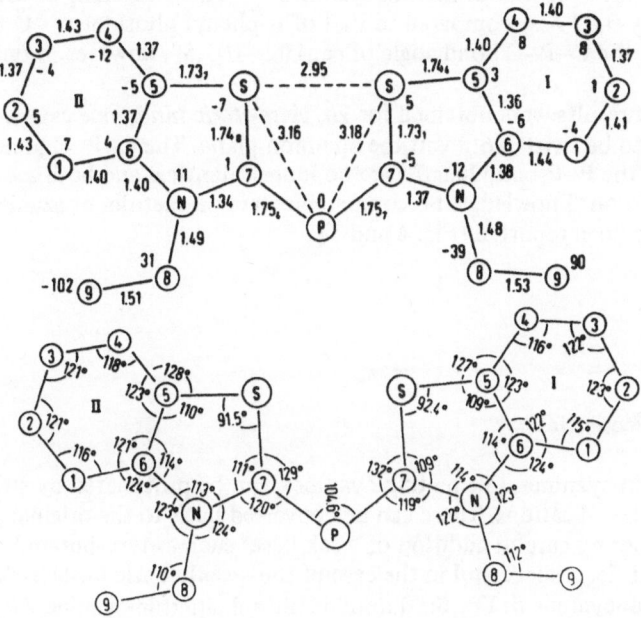

Fig. 4. X-ray structure of bis-[N-ethyl-benzthiazole-(2)]-phospha-methin-cyanine-perchlorate

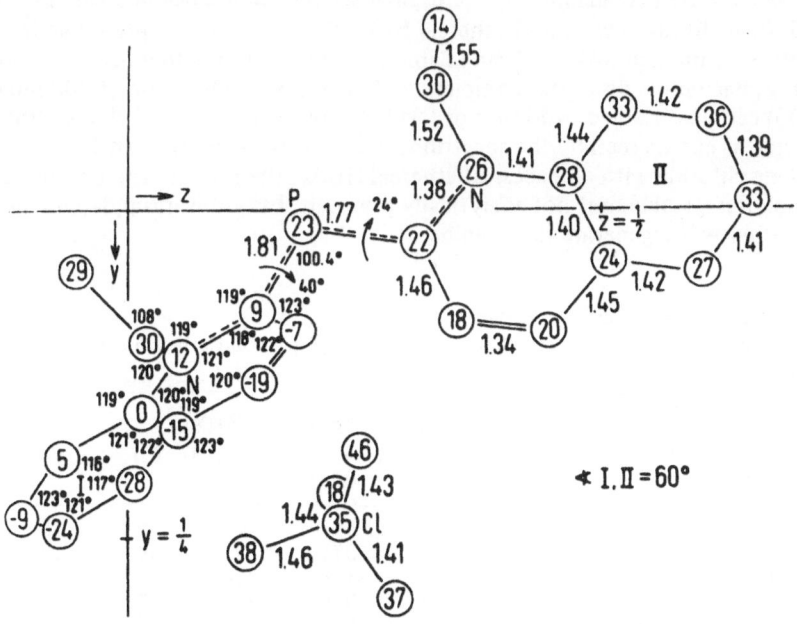

Fig. 5. X-ray structure of bis-[N-ethyl-quinoline-(2)]-phospha-methin-cyanine-perchlorate

17

symmetrical, as in the case of methin-cyanines [18]. The shortening of the two
P–C bonds by 0,07 Å as compared to that of triphenyl-phoshine (P–C + 1,83 Å[19])
is significant; the C–P–C bond angle of ca. 102–103,5°, however, is only a little
wider (1,5°).

Different results were obtained for *7b*. Here *steric hindrance* causes the two
ring systems to be twisted 60° out the common plane. The C–P–C bond angle is
only 100,4°, the P–C bond lengths are no longer identical and both are longer
than in *6b*. To our knowledge, no comparable data on methin- or azamethin-
cyanines have been reported (Fig. 4 and 5).

D. Chemical Properties

Phosphamethin-cyanines, like methin-cyanines, can be protonated by strong acids,
forming colorless dications. These can be converted back to the original phosphamethin-cyanines by careful addition of weak bases such as tert.-butanol. This acid-
base reaction is least successful in the case of the weakly basic bis-benzthiazole-
phosphamethin-cyanine *6*. For bis-quinoline-phosphamethin-cyanine *7 b*, we
obtained with perchloric acid in glacial acetic acid the absorption spectrum of
the N-ethyl-quinolinium salt.

The more basic benzimidazole-phosphamethin-cyanines can be easily protonated. In acetonitrile with 50% ethereal HBF_4 the cyanine absorption bands
at 440–421 nm and 347–335 nm (Table 1) of bis-[1,3-dimethyl-benzimidazole-
2]-phosphacyanine disappear completely. A new sharp absorption at 300 nm (ϵ =
30000) can be observed. Addition of tert-butanol restores 90% of the starting
material, as can be seen by the spectrum. If the protonation is carried out in
ethylene chloride with an excess of ethereal HBF_4, the product can be isolated
as a colorless, stable, but not analytically pure salt. Ethyldiisopropylamine or
more ether will regenerate the cyanine.

Table 3. Benzimidazole-phosphamethincyanines *8*
Absorption maxima of phosphamethin-cyanines *8* in comparison to their Ag-addition compounds

| | R_1 | R_2 | Cyanin | | AgBF$_4$-Komplex | | | |
			λ_{max}-I (nm)	λ_{max}-II (nm)	λ_{max}-I (nm)	λ_{max}-II (nm)	ϵ I	ϵ II
8a	CH_3	CH_3	417	332	367	327	50	5
8d	C_2H_5	C_2H_5	434	340	375	323	59	17
8b	CH_3	C_2H_5	420	333	358	320	62	13
8c	CH_3	$CH(CH_3)_2$	422	338	369	324	53	14
8e	C_2H_5	$CH(CH_3)_2$	437	345	375	327	62	18

By treating benzimidazole-phosphamethin-cyanines *8a* with methanolic silver tetrafluoro-borate solutions, Greif [9] isolated colorless, crystalline *silver complexes*. Their absorption bands lie at shorter wavelengths. The extinction coefficients are little changed. Here, too, a clear dependence of the position of the absorption maxima on the nature of the N-alkyl substituents can be observed (Table 3).

Addition of excess methanol destroys the silver complexes and regenerates the phosphamethin-cyanine salts. It is more advantageous to bind the $\overset{\oplus}{\text{Ag}}$ ions by addition of tris-β-cyanoethyl-phosphine.

Mercuric chloride in methanol also reacts with compounds *8* (in dichloromethane), forming colorless *mercury complexes*, which can in turn be reconverted to the cyanines *8*. Such addition compounds are stable only as solids, decomposing rather quickly in solution. Mercuric acetate in methanol reacts rapidly with the formation of elemental mercury, where by the phosphamethin-cyanines are destroyed; uniform products from this reaction have not as yet been isolated.

The reaction of diazonium salts with methin-[20] and phosphamethin cyanines gives interesting results[11] which are now investigated in detail.

The action of *water* on cyanines *8d* for longer periods of time (with or without careful exclusion of oxygen) leads to 1,3-diethyl-benzimidazolium-tetrafluoroborate *19*. This corresponds to the hydrolytic cleavage of quaternary phosphonium salts.

A similar hydrolytic cleavage, which we initially interpreted as an *oxidative cleavage* [2b], was also observed in the case of bis-benzthiazolphosphamethin-cyanine *6b*, leading to N-ethyl-benzthiazolium perchlorate.

Treatment with H_2O_2 in alkaline solution affords *oxidation products*. For example, *6b* is oxidized to N-ethylbenzthiazolone-(2) *20*.

In summary, the chemistry of the phosphamethin-cyanines as far as it has been investigated to date resembles that of the methincyanines. However, the phosphamethin-cyanines are considerably more reactive. The smooth cleavage of the P–C bond has no counterpart in methincyanine chemistry.

19

III. λ^3-Phosphorins

A. Synthesis

1. Method A: Reaction of Pyrylium Salts with Tris-hydroxymethylphosphine

Our attempts [21], as well as those of C. C. Price [22], to react 2.4.6-triphenylpyrylium salts with phosphine, phenylphosphine or tris-hydroxymethyl-phosphine in the hope of isolating phosphorins or their P-substituted derivatives were unsuccessful. In contrast, Märkl [5], applying essentially the same principle but using *pyridine as base and solvent*, succeeded by heating 2.4.6-triphenylpyrylium-tetrafluoroborate *21* with tris-hydroxymethyl-phosphine[e] *5*. He was able to isolate 2.4.6-triphenyl-λ^3-phosphorin *22*, m. p. 171–172 °C, as the first λ^3-phosphorin in 20–25% yield.

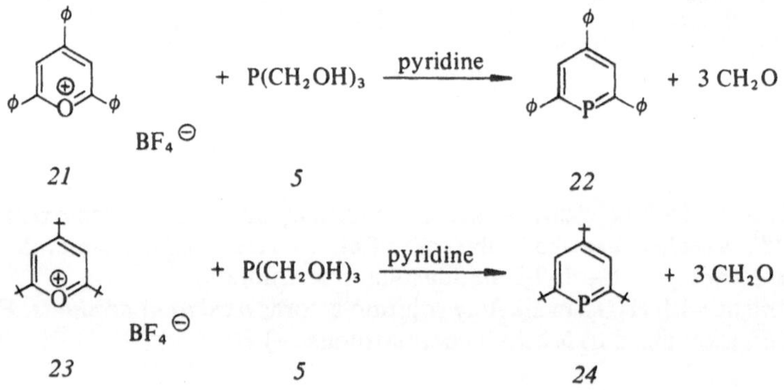

Using this procedure, numerous other aryl-substituted λ^3-phosphorins [23] (see also Table 4), as well as the first alkyl-substituted analog, 2.4.6-tri-tert-butyl-λ^3-phosphorin *24*, were synthesized [24].

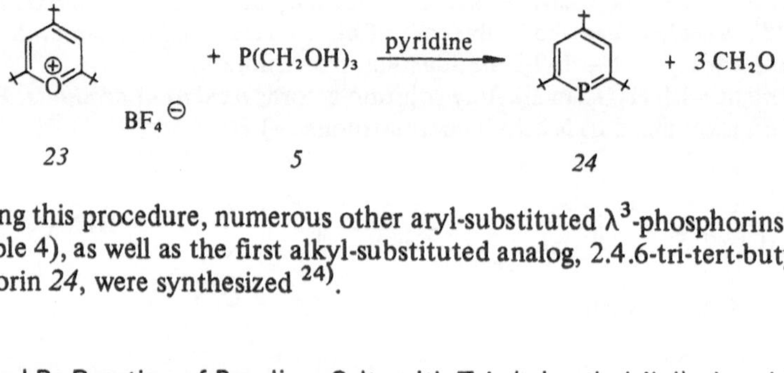

2. Method B: Reaction of Pyrylium Salts with Tris-(trimethylsilyl)-phosphine

This second method is also due to Märkl [25] and again involves pyrylium salts. However, instead of CH_2OH groups, which are easily split off under basic condi-

---

[e] See footnote on p. 7

20

tions -Si(CH$_3$)$_3$ groups are used as protective groups of the phosphine. Since the starting material P[Si(CH$_3$)$_3$]$_3$ is not easily prepared, this method (B) offers no advantages over Method A.

21   22

3. Method C: Reaction of Pyrylium Salts with Phosphoniumiodide

Märkl [26] has developed a third and extremely useful synthetic sequence again reacting pyrylium salts but merely with PH$_4$J or PH$_3$ in the presence of acids. The components are heated with butanol as solvent in a pressure flask or glass autoclave at 120°–130 °C for 24h. The corresponding λ^3-phosphorins are isolated in yields often greater than 50%. This method is particularly useful for the synthesis of methyl substituted λ^3-phosphorins (e. g. 26) where tris-hydroxymethyl-phosphine cannot be used because its relatively large size leads to steric repulsion with the large substituents at C–2 and C–6 positions. The addition thus occurs at the less sterically hindered C–4 position, when ring closure to the phosphorin system is no longer possible. A similar behavior was previously observed in the condensation of nitromethane anions with 4-methyl-2,6-diphenyl-pyrylium salts [27]. Moreover, in the reaction with tris-hydroxymethylphosphine the liberated formaldehyde probably condenses with the methyl groups of the pyrylium salt, yielding undesired side products.

25   26

The reactions in Methods A, B and C, which all start from pyrylium salts [28], are analogous to the well-known conversions of 2.4.6-substituted pyrylium salts 27 with ammonia, primary amines, hydrogen sulfide or the anions of CH activated compounds to the corresponding heterocyclic or isocyclic aromatic systems [29]. The first step involves addition of the basic phosphine at C–2 (or C–6) to form 28. Ring-opening, ring-closure and elimination of water are likely steps in the formation of the product 2.

21

However, radical intermediates cannot be definitely exclucled, at least not in the reaction of pyrylium salts in pyridine. *Steuber* [30] showed that such pyrylium salts as 2.4.6-triphenyl-pyrylium-or 2.4.6-tri-tert-butyl-pyrylium-tetrafluoroborate can be reduced to stable pyryl radicals *32* by pyridine; this reduction proceeds particularly smoothly if traces of copper powder are added.

At the same time the pyridine is oxidized to a pyridine radical cation which by dimerization and proton loss forms 4,4'-dipyridyl. However, if 2.4.6-triphenylpyridine *31* is used instead of pyridine, it forms the stable radical cation *33*, which can be observed in the ESR-spectrum in addition to the pyryl radical *32* [31].

These three methods can be used to prepare a large number of 2.4.6-tri-or higher substituted λ^3-phosphorins (Table 4). *Märkl* [32] describes only a single case (no details given) in which a pyrylium salt with an unsubstituted α-position 2.4.5-triphenylpyrylium salt is used to form a λ^3-phosphorin which has no substituent at the 6-position: 2.4.5-triphenyl-λ^3-phosphorin (Table 2, no. 27).

4. Method D: Elimination of HCl from Cyclic Phosphine Chlorides

An entirely different method for the preparation of λ^3-phosphorins utilizes phosphine chlorides *36* or *40*. Treatment with bases leads to HCl elimination and to the formation of the λ^3-phosphorins *37* and *41*, respectively. The phosphine chlorides can be prepared from cyclic phosphinic acids *34* by reduction with diphenylsilane to the cyclic phosphines *35* followed by chlorination with phosgene. Alternatively, suitable bromides, such as *38* or *39*, can be reacted with Grignard reagents followed by treatment with diethylamino-dichlorphosphane:

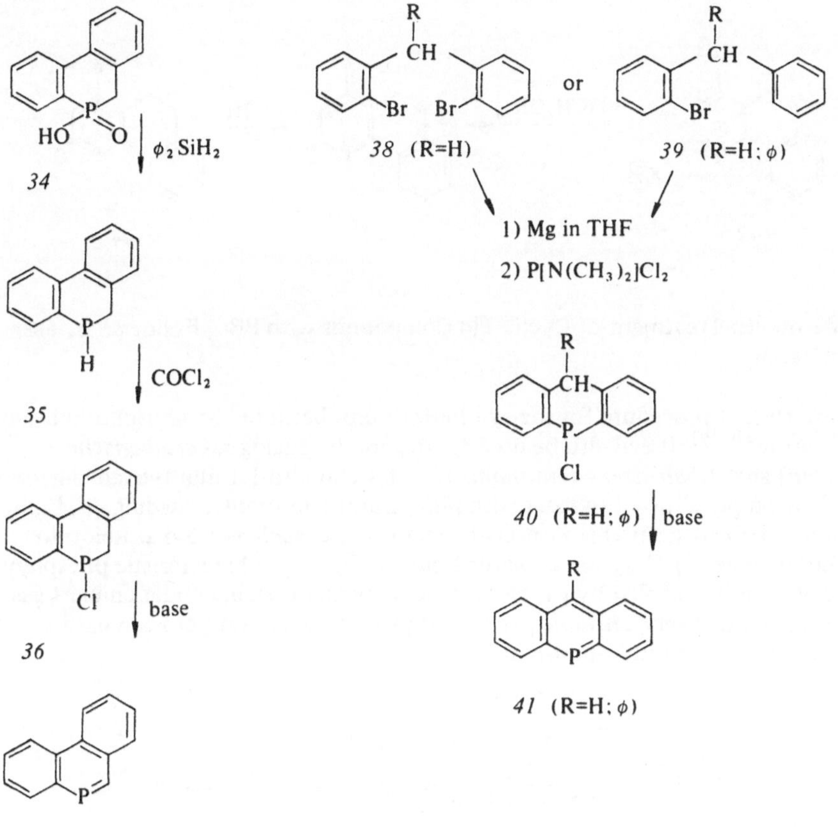

In this manner *de Koe, van Veen and Bickelhaupt* [33, 34, 35] were able to prepare the very unstable 9-phosphaphenanthrene *37* [f] as well as 9-phospha-anthracene *41a*. These compounds were stored in solution and could not be isolated as solids. However, if the C−10 position of *41* is blocked by a phenyl group, the same procedure leads to the much more stable yellow crystalline, 10-phenyl-9-phospha-anthracene *41b* (4-phenyl-dibenzo[b, e]-phosphorin) (m. p. 173−176 °C). Another stable substituted phosphaphenanthrene *43* has been prepared (*albeit* in

---

[f] Meanwhile P. de Koe could synthesize by a similar procedure cristalline air sensitive 10-Phenyl-9-phospha-phenanthrene m.p. 124−131 °C, λ_{max} 339 nm (11900) in ether; (private communication by Dr. Vermeer).

23

low yields) according to Method A by treatment of the appropriate pyrylium salt *42* with tris-hydroxymethylphosphine [36]. All attempts to dehydrogenate tetra-hydro-naptho-λ^3-phosphorin *44*, (easily prepared from the corresponding pyrylium salt by Method A) to *43* failed [37].

42   *43*   *44*

5. Method E: Treatment of Cyclic Tin Compounds with PBr₃ Followed by HBr Elimination

A very elegant procedure for the synthesis of unsubstituted λ^3-phosphorin is due to A. Ashe [38,39]. It can also be used to prepare the analogous *arsabenzene* (*arsenin*) and *stibabenzene* (*antimonin*) [59]. 1,4-Dihydro-1,1-dibutylstanna-benzene *45* [40] is simply allowed to react with phosphorus-tribromide, yielding *46*. Elimination of HBr by a suitable proton-abstracting base, such as 1,5-diaza-bicyclo [4.3.0] nonene-(5) [41], gives a volatile liquid *47* having a characteristic phosphine odor. *47* can be purified by gas chromatography and is stable under an inert gas atmosphere; it is very *unstable in air*. This procedure has not yet been used to synthesize substituted λ^3-phosphorins.

45   *46*   *47*

6. Method F: Elimination of the 1.1-Substituents in λ^5-Phosphorins

The direct synthesis of 1.1-substituted λ^5-phosphorins requires many tedious steps [6] (see p. 76). Cleavage of the 1.1-substituents is possible in some cases (see p. 90), but on the whole this method has no preparative value.

On the other hand, λ^3-phosphorins can easily be converted to 1.1-hetero-λ^5-phosphorins which can be reconverted by cleaving the 1.1-substituents to the λ^3-phosphorins (see p. 88). The value of this procedure lies in the fact that the phosphorus of the λ^3-phosphorins can be protected by the 1.1-hetero groups.

24

Functional groups can than be introduced or altered. The new λ^5-phosphorin-compound is then reconverted to a λ^3-phosphorin which cannot be synthesized in any other way.

7. Method G: 4.4'-Bis-λ^3-phosphorins from Thermolysis of 1.4-Dihydro-λ^3-phosphorins

Märkl, Fischer and Olbrich [48] showed that thermolysis of 1.1'-dibenzyl-1.1'-diphospha-4.4'-pyrylene at 350 °C under careful exclusion of oxygen leads to 4.4'-bis-2.6-diphenyl-λ^3-phosphorin of m. p. 236°−239 °C.

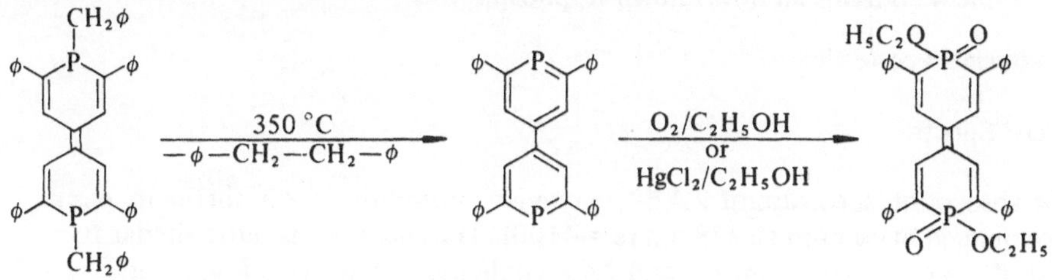

The UV spectrum ($\lambda_{max_1} = 328$ nm, $\epsilon_1 = 16600$, $\lambda_{max_2} = 280$ nm, $\epsilon_2 = 64500$) is similar to that of 4.4'-bis-2.6-diphenylpyridine ($\lambda_{max_1} = 317$ nm, $\epsilon_1 = 16700$, $\lambda_{max_2} = 246$ nm, $\epsilon_2 = 79500$). The compound is easily oxidized; oxidation in air at 220 °C yields a *deep red polymer*, but oxidation in the presence of ethanol and air or mercuric chloride gives the bis-phosphinic acid ester (m. p. 318–320 °C) which has UV absorption bands at 440 nm (55800) and 317 nm (26300).

Märkl and coworker have developed the following route to 1.1-dibenzyl-1.1'-diphospha-4.4'-pyrylene:

25

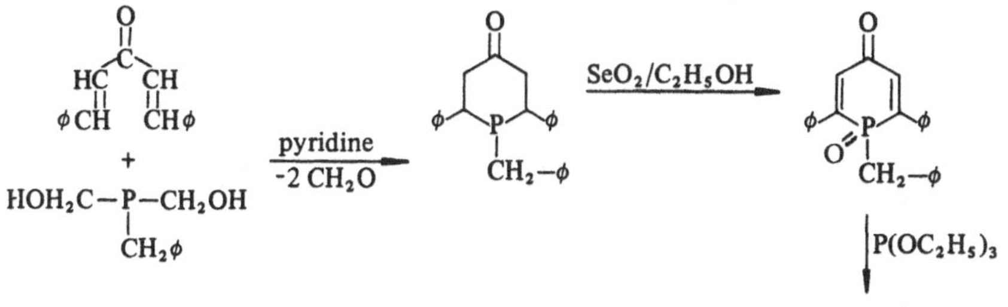

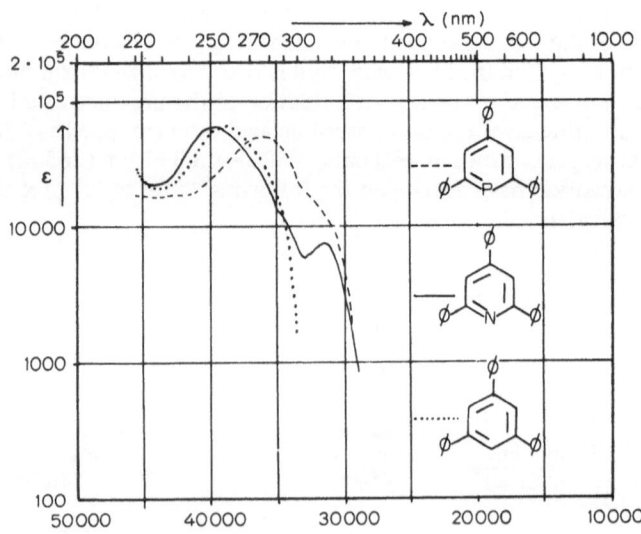

The direct reaction of 4.4′-bis-2.6.2.′6′-tetraphenyl-pyrylium tetrafluoroborate with tris-hydroxymethylphosphine did not lead to the desired product.

Table 4 contains all now known λ^3-phosphorins.

B. Physical Properties

1. UV Spectra

The absorption spectrum of 2.4.6-triphenyl-λ^3-phosphorin *22* in methanol has a pronounced maximum at 278 nm (ϵ = 4100). The spectrum is quite similar to that of 2.4.6-triphenylbenzene and 2.4.6-triphenylpyridine (see Fig. 6 and Table 5).

Fig. 6. UV spectra of 2.4.6-triphenyl-λ^3-phosphorin, -pyridine and -benzene in cyclohexane

26

Table 4. λ^3-Phosphorins

$$\begin{array}{c} R^4 \\ R^3 \diagdown \diagup R^5 \\ R^2 \diagdown \diagup R^6 \\ P \end{array}$$

	R²	R³	R⁴	R⁵	R⁶	m.p. [°C]	Method	Lit.
1	H	H	H	H	H	—	E	38)
2	CH₃	H	CH₃	H	CH₃	135/12(b. p.)	C	32)
3	CH₃	H	C₆H₅	H	CH₃	62–63	C	42)
4	CH₃	H	C₆H₅	H	C₆H₅	79–81	C	26)
5	C₆H₅	H	CH₃	H	C₆H₅	118–120	C	26)
6	C₆H₅	H	C₂H₅	H	C₆H₅	65–66	C	43)
7	C₆H₅	H	CH(CH₃)₂	H	C₆H₅	oil	C	43)
8	C₆H₅	H	CH₂C₆H₅	H	C₆H₅	97	C	43)
9	C(CH₃)₃	H	C(CH₃)₃	H	C(CH₃)₃	88	A	24)
10	C(CH₃)₃	H	C₆H₅	H	C(CH₃)₃	104–5	A	23)
11	C(CH₃)₃	H	C₆H₄OCH₃(4)	H	C(CH₃)₃	116–116.5	A	23)
12	C(CH₃)₃	H	C₆H₄OH(4)	H	C(CH₃)₃	141	from 11	44)
13	C(CH₃)₃	H	C₆H₄-O-COCH₃	H	C(CH₃)₃	127	from 12	44)
14	C(CH₃)₃	H	C₆H₄OCH₃(2)	H	C(CH₃)₃	oil	A	44)
15	C₆H₅	H	C(CH₃)₃	H	C(CH₃)₃	87.5–88	A	23)
16	C(CH₃)₃	H	C₆H₅	H	C(CH₃)₃	—	A	37)
17	C(CH₃)₃	H	C₆H₄CH₃(4)	H	C(CH₃)₃	99–101	A	37)
18	C(CH₃)₃	H	C₆H₄CH₃(4)	H	C(CH₃)₃	160–163	A	37)
19	C(CH₃)₃	H	C₆H₄OCH₃(4)	H	C(CH₃)₃	—	A	37)
20	C(CH₃)₃	H	C₆H₄Cl(4)	H	C(CH₃)₃	150–2	A	37)
21	C(CH₃)₃	H	C₆H₅	H	C(CH₃)₃	135–136	A	36)
22	[2-methylphenethyl group]	H	C₆H₅	H	[2-methylphenethyl group]	193–7	A	37)
23		H	C₆H₄CH₃(4)	H		178–81	A	37)
24		H	C₆H₄OCH₃(4)	H		204–209	A	37)
25		H	C₆H₄Cl(4)	H		194–9	A	37)

27

Table 4 (continued)

	R²	R³	R⁴	R⁵	R⁶	m. p. [°C]	Method	Lit.
26	(ring)	H				unstable	D	33)
27	(ring)		C_6H_5	(ring)	(ring)	173–6	D	35)
28	H	(ring)				unstable	D	34)
29	H	C_6H_5	C_6H_5	H	C_6H_5	–	A, B, C	32)
30	C_6H_5	H	C_6H_5	H	C_6H_5	171–172	A	5, 25, 26)
31	C_6D_5	H	C_6H_5	H	C_6D_5	167	A	23)
32	C_6D_5	H	C_6D_5	H	C_6D_5	168–71	A	23)
33	C_6H_5	C_6H_5	C_6H_5	H	C_6H_5	209–10	A	25, 26)
						188–189.5	A	9)
34	C_6H_5	C_6H_5	C_6H_5	C_6H_5	C_6H_5	253–254	A, B	25)
						216–217	A	9)
35	$C_6H_4CH_3(4)$	H	C_6H_5	H	C_6H_5	155–156.5	A	23)
36	$C_6H_4CH_3(4)$	H	C_6H_5	H	$C_6H_4CH_3(4)$	133–134	A	25)
37	$C_6H_4CH_3(4)$	H	$C_6H_4CH_3(4)$	H	$C_6H_4CH_3(4)$	167–170	A	37)
38	$C_6H_3[C(CH_3)_3]_2$ (2,4)	H	C_6H_5	H	$C_6H_3[C(CH_3)_3]_2$ (2,4)	220	A	45)
39	α-Naphthyl	H	C_6H_5	H	C_6H_5	163–4	A	23)
40	$C_6H_4OCH_3(4)$	H	C_6H_5	H	C_6H_5	161.5–3	A	23)

Table 4 (continued)

	R²	R³	R⁴	R⁵	R⁶	m. p. [°C]	Method	Lit.
41	C_6H_5	H	$C_6H_4OCH_3(4)$	H	C_6H_5	106 (110–112)	A	25, 23) 46)
42	$C_6H_4OCH_3(4)$	H	$C_6H_4OCH_3(4)$	H	C_6H_5	134–6	A	23)
43	$C_6H_4OCH_3(4)$	H	C_6H_5	H	$C_6H_4OCH_3(4)$	136–7 132–3	A	25) 23)
44	$C_6H_4OCH_3(4)$	H	$C_6H_4OCH_3(4)$	H	$C_6H_4OCH_3(4)$	105–6	A	25)
45	$C_6H_4OCH_3(4)$	H	$C_6H_4C_6H_5(4)$	H	C_6H_5	148.5–50	A	23)
46	$C_6H_4Cl(4)$	H	C_6H_5	H	C_6H_5	166–7	A	23)
47	$C_6H_4Cl(4)$	H	$C_6H_4Cl(4)$	H	$C_6H_4Cl(4)$	181–2	A	23)
48	C_6H_5	H	$C_6H_4Br(4)$	H	C_6H_5	148–9	A	47)
49	C_6H_5	H	$C_6H_4N(CH_3)_2(4)$	H	C_6H_5	116–17	A	46)
50	C_6H_5	H	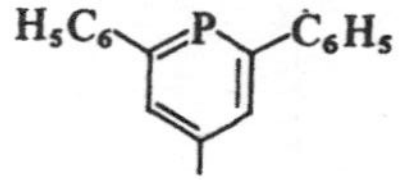 (structure: H_5C_6, C_6H_5, P)	H	C_6H_5	236–8	g	48)
51	C_6H_5	H	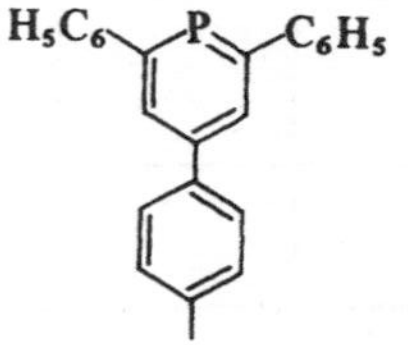 (structure: H_5C_6, C_6H_5, P)	H	C_6H_5	218	A, B, C	48)

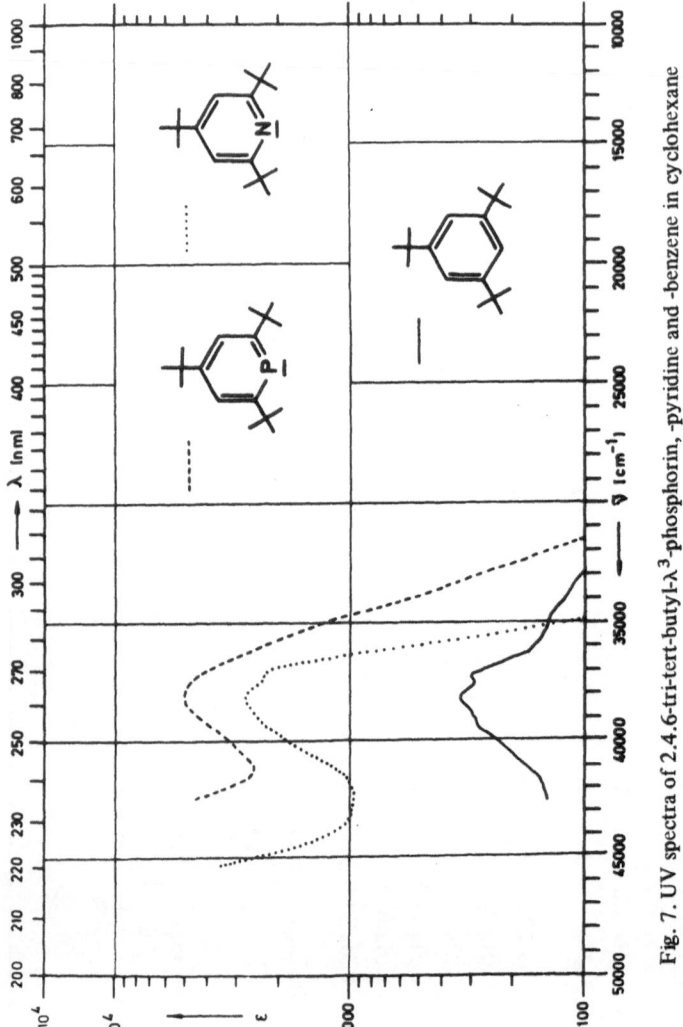

Fig. 7. UV spectra of 2.4.6-tri-tert-butyl-λ^3-phosphorin, -pyridine and -benzene in cyclohexane

The spectral properties of the λ^3-phosphorin ring system can be identified more easily in molecules with alkyl substituents such as 2.4.6-tri-tert-butyl-λ^3-phosphorin *24* (Fig. 7) or 2.4.6-trimethyl-λ^3-phosphorin. Märkl [49] attributes the long-wave absorption (shoulder at 312 nm) to an $n \rightarrow \pi^*$ transition (see, however, p. 38), the absorption in the center to an $^1L(a)$-transition, and the short-wave absorption to an $^1L(p)$-transition. The values are listed in Table 5 together with the absorption bands of the unsubstituted λ^3-phosphorin.

Table 5. Absorption bands of identically substituted benzene, pyridine, pyrylium and λ^3-phosphorin compounds

Substituents in 2.4.6-position	Benzene λ_{max} [nm]		Pyridine λ_{max} [nm]		Pyrylium-Salt λ_{max} [nm]		λ^3-Phosphorin λ_{max} [nm]		Lit.
C_6H_5	254	56000	254	49500	355	22000	278	41000	5,9
			312	9390	403	16400	314	12600	
$C(CH_3)_3$	262	333	262	2800	240	2670	262	4100	24
					290	7800			
CH_3	213	8200	212	8800			227	27200	
	263	219	265	6630			261	7390	49
							312	246	
H	180	25000	175	80000			213	19000	38
	193–204	8000	192	6300			246	8500	
	255	250	250	2000					
	(230–270)		270	450					

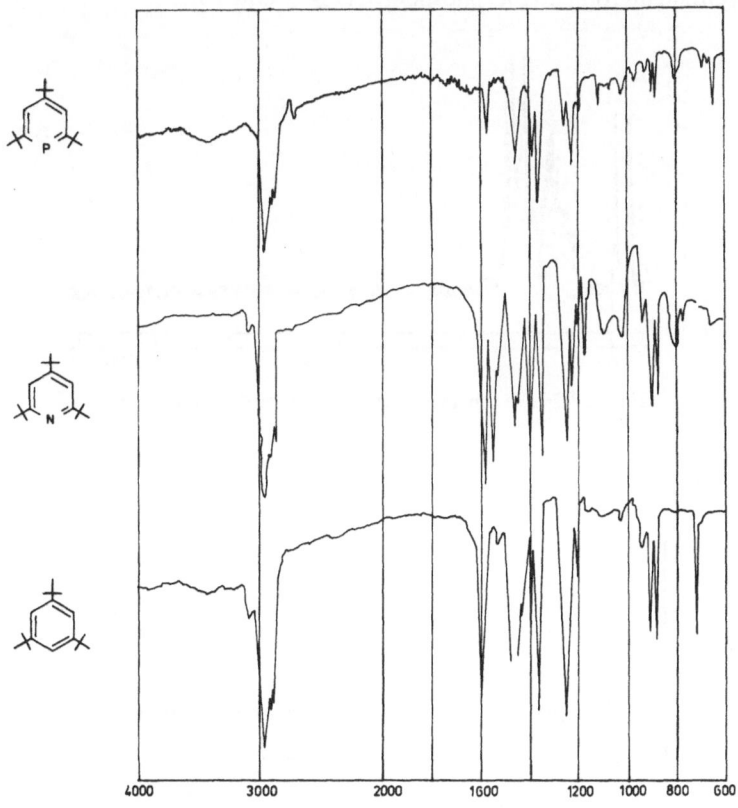

Fig. 8. IR spectra of 2.4.6-tri-tert-butyl-λ^3-phosphorin, -pyridine and -benzene (in KBr)

2. IR Spectra

The IR spectra of the corresponding benzene, pyridine and λ^3-phosphorin derivatives resemble each other, although λ^3-phosphorin is spectrally closer to benzene than to pyridine. Only in the region $1200-1400$ cm^{-1} additional bands are to be detected. However, these alone do not suffice to identify the λ^3-phosphorin system. Fig. 8 shows the IR spectra of 2.4.6-tri-tert-butyl-λ^3-phosphorin compared with those of the anologous benzene and pyridine derivatives.

3. NMR Spectra

a) The ^{1}H-NMR spectrum of 2.4.6-triphenyl-λ^3-phosphorin shows an absorption band at $\delta = 7-7,8$ ppm due to the 15 protons of the three phenyl groups. More significant are the two peaks (2 protons) centered at $\delta = 8,1$ ppm (in CDCl$_3$ TMS internal standard) with a coupling constant of 6 Hz. These signals are due to the two equivalent protons at C-3 and C-5 which are split into a doublet by phosphorus, as can be seen by inspection of the NMR spectrum of tris-pentadeuterophenyl-λ^3-phosphorin (Fig. 9 and 10).

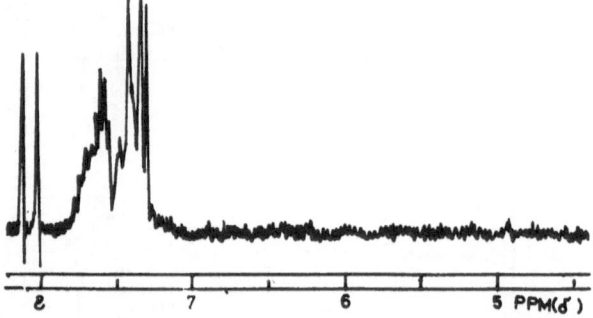

Fig. 9. ^{1}H-NMR spectrum of 2.4.6-λ^3-triphenylphosphorin *22*

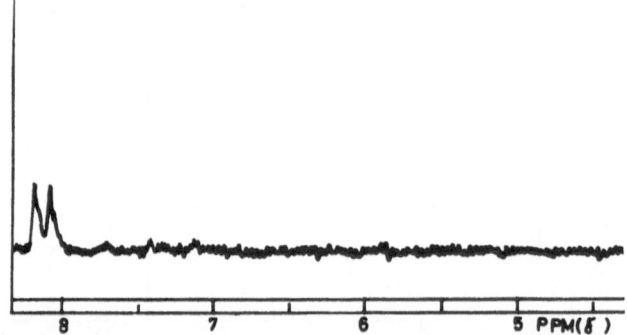

Fig. 10. ^{1}H-NMR spectrum of 2.4.6-tris-pentadeutero-phenyl-λ^3-phosphorin

λ^3-Phosphorins are thus *"aromatic"* in the sense that they show a clear *ring-current effect*. The spectrum of 2.4.6-tri-tert-butyl-λ^3-phosphorin *24* in CDCl$_3$ is quite similar in that the two ring protons absorb at δ = 7,77 ppm with a coupling constant of $J_{P-C-C-H}$ = 6 Hz. The 2.6-tert-butyl groups can be found at δ = 1,47 ppm (18 H), and 4-tert-butyl group at δ = 1,37 ppm (9H); neither shows any appreciable coupling with phosphorus.

For comparison, Table 6 contains some NMR data of analogous aromatic compounds.

Table 6. ^{1}H-NMR spectral data of identically substituted benzene, pyridine, pyrylium and λ^3-phosphorin compounds (δ in ppm, TMS internal standard)

Substituents in 2.4.6- positions	Benzene	Pyridine	Pyrylium Salt	λ^3-Phosphorin
C$_6$H$_5$	7.3–7.8 (m, 17H)	7.8 (s, 2H) 7.3–7.6; 8.0–8.3 (m, 15H)	8.42 (s, 2H) 7.5–8.4 (m, 15H)	8.1 (d, 2H, J = 6Hz) 7–7.8 (m, 15H)
C(CH$_3$)$_3$	7.20 (s, 2H) 1.34 (s, 27H)	7.02 (s, 2H) 1.34 (s, 18H) 1.30 (s, 9H)	7.78 (s, 2H) 1.51 (s, 18H) 1.47 (s, 9H)	7.77 (d, 2H, J = 6Hz) 1.46 (d, 18H, J = 2Hz) 1.37 (s, 9H)
H	7.26 (s)	8.48 (α, 5.5 Hz)	9.55 (m, 2H) 9.2 (m, 1H) 8.4 (m, 2H) [in (CF$_2$Cl)$_2$ C(OH)$_2$]	8.6 (J_1 = 38Hz J_2 = 10Hz) 7.25–8.4 (m)

Thiopyrylium-tetrafluoroborate has the following absorptions ((CF$_2$Cl)$_2$C(OH)$_2$ solvent): δ = 10,0 ppm (m, 2H); δ = 8,9 ppm (m, 3H); 2.4.6-tri-tert-butyl-pyridinium-tetra-fluoroborate (C$_2$H$_5$Cl$_2$ solvent): δ = 7,88 ppm (d, 2H, J = 2Hz); 1,63 ppm (s, 18H); 1,47 ppm (s, 9H). The signal for the N–H proton is not observed.

Unsymmetrically 2.4.6-λ^3-phosphorins with different substituents at the C–2 and C–6 positions can easily be distinguished from the symmetric compounds, since the C–3 and C–5 protons couple with each other, giving rise to an ABX splitting pattern ($J_{H,H}$ = 1–2 Hz; $J_{H-C-C-P}$ = 6 Hz). For example,

49

33

2-tert-butyl-4-(4'-methoxyphenyl)-6-phenyl-λ^3-phosphorin *49* gives rise to 8 peaks (δ = 8,10 to 7,88 ppm) for the *a* and *a'* protons. The 9 protons of the two benzene rings appear as multiplets between δ = 7,7 and 6,85 ppm, the 3 protons of the methoxy group as a singlet at δ = 3,70 ppm, and the 9 protons of the tert-butyl group as a singlet with a very small splitting by phosphorus at δ = 1,55 ppm.

b) The ^{31}P–NMR signals of both the aryl and alkyl substituted λ^3-phosphorins appear between δ = –170 and –180 ppm (85% H_3PO_4, external standard). They are thus quite unlike those of the phosphamethin-cyanines and appear to be rather characteristic of λ^3-phosphorins. The small coupling with the protons is usually not easily detected. Some characteristic values are given below:

2.4.6-triphenyl-λ^3-phosphorin $\qquad$ ^{31}P: δ = –178,2 ppm

2.4.6-tri-tert-butyl-λ^3-phosphorin $\qquad$ ^{31}P: δ = –178,5 ppm

4. Mass Spectra

The parent peak, usually having the highest m/e value, can easily be identified in all cases. Triphenyl-λ^3-phosphorin also has an intense peak at m/e = 120, which Märkl [32] attributes to the C_6H_5–C $\equiv$ P fragment. Fig. 11 shows the mass spectra of 2.4.6-tri-tert-butyl-λ^3-phosphorin *24*. The parent peak as well as the fragments CH_3 (15) and $C(CH_3)_3$ (57) are easily identified.

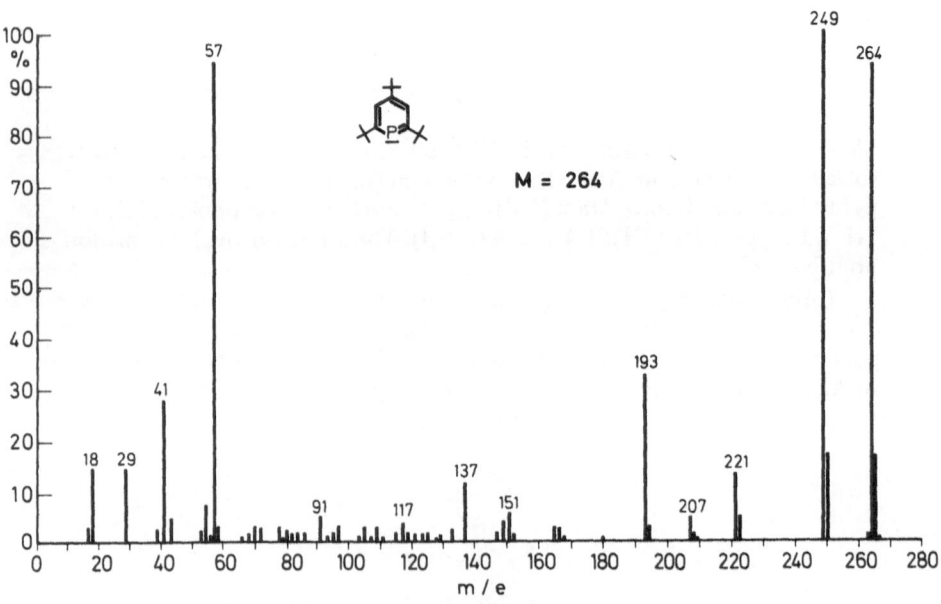

Fig. 11. Mass spectrum of 2.4.6-tri-tert-butyl-λ^3-phosphorin

5. X-Ray Analysis

X-ray structure determinations of *50* and *51* were reported simultaneously by Bart and Daly [42] and by Fischer, Hellner, Chatzidakis and Dimroth [50].

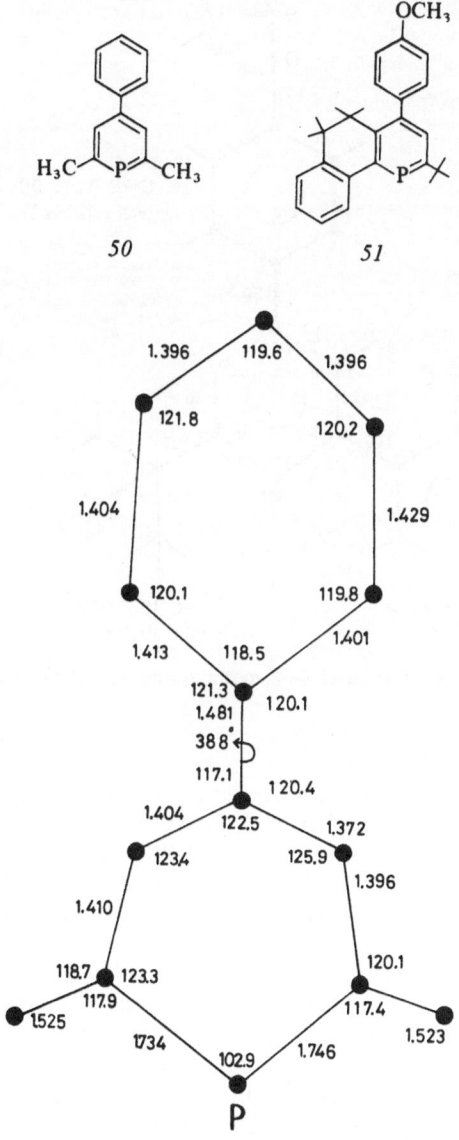

Fig. 12. Crystal structure of 2.6-dimethyl-4-phenyl-λ^3-phosphorin *50* [42]

Figs. 12 and 13 show that the bond distances in the λ^3-phosphorin ring, particularly those of the C—P bonds are nearly equal; the C—P—C angle and the nearly planar shape of the λ^3-phosphorin ring are quite similar in the two systems.

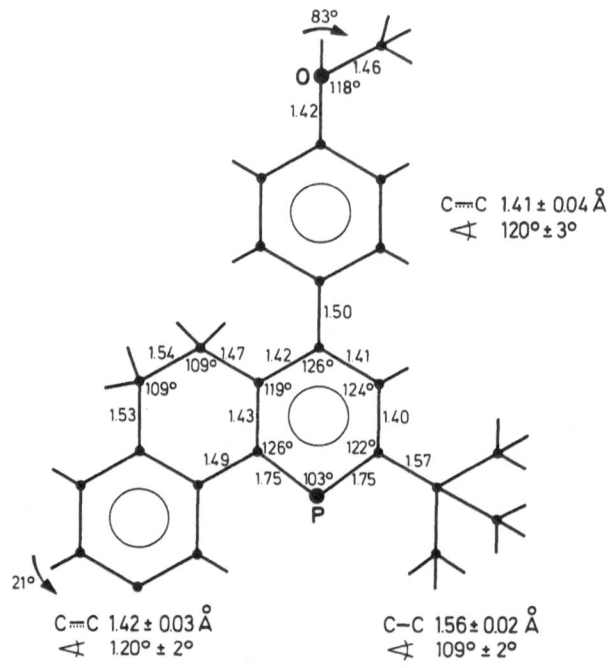

Fig. 13. Crystal structure of 2-tert-butyl-4-(4'-methoxyphenyl)-5.6-dihydronaptho [1.2-b]-λ^3-phosphorin 51[50])

These structure determinations prove that the bonds in λ^3-phosphorin do not alternate (switch), but are rather delocalized over the entire system, much like the aromatic pyridine system. It is noteworthy that the two P—C bond lengths in 50 and also in 51 are equally large, 1,75 and 1,73 Å, and 1,75 and 1,75 Å, respectively. These values are very similar to these of the phosphamethincyanines(p. 16). They lie between those of P—C single bonds (e. g. 1,83 Å in tri-phenylphosphine [51]) and those of the P = C bonds of phosphorylenes (1,65 or 1,68 Å [52]). The C—P—C bond angle has the same value (103°) in 50 and 51 and corresponds to the C—P—C angle found for triphenylphosphine (also 103°) [51]. The ring in the λ^3-phosphorins is practically planar and the C—C bond lengths are much like those of phenyl rings; however, the C—C—C bond angles are slightly wider. In summary, the shapes of the λ^3-phosphorin ring system are much like those of the iso- and heterocylic analogs. The only significant difference lies in the fact that the large P atom causes some widening in the C—C—C bond angles.

6. Photoelectron Spectra

The photoelectron (PE) spectra of 2.4.6-tri-tert-butyl-λ^3-phosphorin *24* and 2.4.6-tri-tert-butylpyridin *53* have been recorded by Oehling, Schäfer and Schweig [53] (Fig. 14).

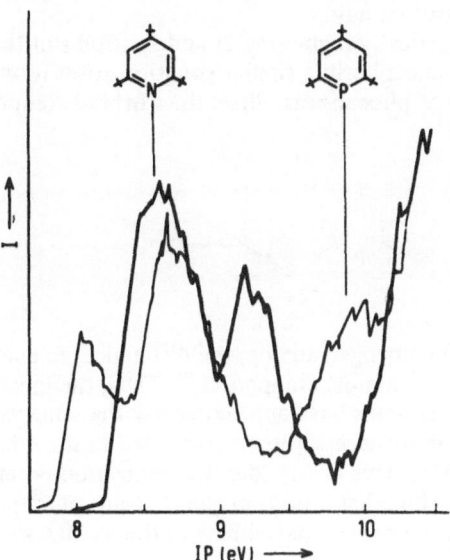

Fig. 14. Photoelectron spectra of 2.4.6-tri-tert-butyl-pyridine *53* and 2.4.6-tri-tert-butyl-λ^3-phosphorin *24*

The λ^3-phosphorin is ionized more easily (maxima at 8,0 and 8,6 eV) than the pyridine analog (maxima at 8,6 and 9,3 eV) [g].

C. Bonding in the λ^3-Phosphorin System

On the basis of CNDO/2 calculations on the model compounds 2.4.6-trimethyl-pyridine and 2.4.6-tri-methyl-λ^3-phosphorin, Schweig and coworkers [53, 54] found that the MO sequences of pyridine and λ^3-phosphorin do *not* correspond to each other. For simplicity, methyl- rather than tert-butyl groups were used in the calculations (whereas the PE and UV spectra of the tert-butyl compounds have been recorded; the synthesis of the unsubstituted λ^3-phosphorin was unknown at the

---

[g] The tert-butyl groups cause the IP of pyridine to be lowered by ca. 1,0–1,2 eV. A similar effect should operate in the λ^3-phosphorin system.

time). In pyridine the highest occupied MO corresponds to the n orbital of the lone electron pair at the nitrogen atom. In contrast, in the λ^3-phosphorin the highest occupied MO is a π orbital. The calculations point to a $[n, \pi_1, \pi_2, \sigma_1]$ sequence for pyridine and a $[\pi_2, n, \sigma_1, \pi_1]$ sequence for λ^3-phosphorin. Since CNDO/2 calculations on unsaturated systems often give energy values for σ orbitals which are too high, Schweig contends that for λ^3-phosphorin the sequence $\pi_2, n, \pi_1, \sigma_1$ is more reasonable. According to the calculations the n MOs of both systems have the same energy.

In pyridine the energetically high-lying $2s$ and $2p_x$ AO's of the N atom mix to form the n MO. In λ^3-phosphorin a similar situation arises if one mixes the high-lying $3s$ and $3p_x$ orbitals of phosphorus. Since the s orbital component is greater

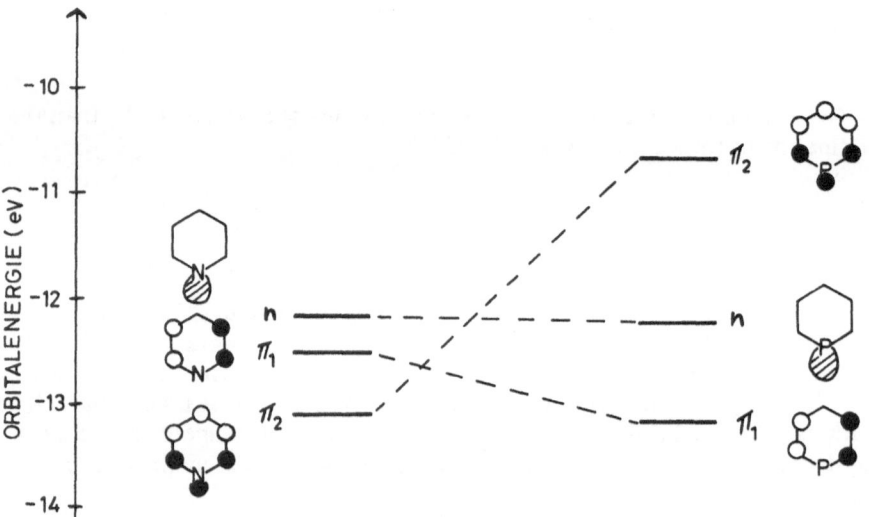

in this case than it is in the nitrogen analog, the differences in energy between the 2 and 3 shells of N and P are almost compensated. Thus, the ionization potentials of the n MO's of the two systems have approximately the same value. The peaks at 8,6 eV in the PE spectra must necessarily be attributed to the n MO's. Similarly, the n MO's of PH_3 and NH_3 have nearly identical ionization potentials. In pyridine the highest occupied π_1 orbital has a node at the N atom. In λ^3-phosphorin however, the $3d_{yz}$AO of phosphorus can participate in the π_1 MO, so *that π conjugation*

Fig. 15. Orbital schemes of pyridine and λ^3-phosphorin

is transmitted through the P atom. Thus, the energy of this π_1 MO in λ^3-phosphorin is somewhat lower than the corresponding π_1 MO of pyridine.

On the other hand, the λ^3-phosphorin π_2 orbital (which in pyridine has a high electron density at the N atom) rises sharply in energy, since the $3p\pi$-$2p\pi$ conjugation between phosphorus and the neighboring carbon atom is much weaker than the corresponding $2p\pi$-$2p\pi$ conjugation in pyridine. This causes the π_2 MO in λ^3-phosphorin to be the highest occupied MO.

The decrease in effective conjugation discussed above is much more important than any possible participation of phosphorus $3d_{xz}$ orbitals. Such weak d-orbital conjugation is not expected to lower the energy of the π_2 MO to any significant degree. Fig. 15 roughly summarizes the orbital schemes of the two heterosystems.

D. Chemical Properties

No chemical studies have been carried out on the unsubstituted λ^3-phosphorin. This highly reactive compound probably does not lend itself to specific chemical transformations. Most likely, the situation here is quite similar to that of pyrylium compounds, where the unsubstituted pyrylium salt [55], unlike the 2.4.6-substituted form, has no preparative significance [29]. In the following discussion we therefore limit ourselves to the reactions of 2.4.6-substituted λ^3-phosphorins.

1. Basicity, π and σ Complexes with Transition Metals and Charge-Transfer Complexes

In contrast to pyridine derivatives, aryl- and alkyl-substituted λ^3-phosphorins cannot be protonated by strong, non-oxidizing acids such as trifluoroacetic acid. Addition of trifluoroacetic acid to cyclohexane solutions of various λ^3-phosphorins fails to produce any change in the UV spectra [47]. Similarly, alkylation by such strong agents as oxonium salts or acylation by acylchlorides cannot be induced at the P atom or any ring C atom. This behavior has also been discussed theoretically [55a].

Addition of a few drops of 60% perchloric or conc. sulfuric acid to a solution of 2.4.6-triphenyl-λ^3-phosphorin *22* affords a deep blue compound which is soluble in polar solvents. This new compound is not the protonated form of the λ^3-phosphorin, but rather a cation which results from oxidation of the λ^3-phosphorin (see p. 50).

According to the acid-base concept of Pearson, λ^3-phosphorins can be viewed as "soft bases"; the lone electron pair at phosphorus is much more delocalized than the lone pair at nitrogen in pyridine. Thus, such soft Lewis acids as Hg^{2+} ions are more likely to react with λ^3-phosphorins (see p. 84).

Investigations using diborane and boronhalides as Lewis acids have not been completed. According to preliminary results of Nöth and Deberitz [56], the vapor pressure of B_2H_6 at low temperatures is reduced significantly upon addition of 2.4.6-triphenyl-λ^3-phosphorin. By raising the temperature the postulated equilibrium *22* $\rightleftharpoons$ *54* can be shifted to the left.

2 [structure 22] + B$_2$H$_6$ ⇌ 2 [structure 54]

22   *54*

Deberitz and Nöth [57)] have also found that 2.4.6-triphenyl-λ^3-phosphorin *22* (in contrast to the sterically hindered 2.4.6-tri-tert-butyl-λ^3-phosphorin) reacts with chromiumhexacarbonyl, expelling CO and forming a σ complex *55*. In refluxing dibutylether both the tri-tert-butyl- and the triphenyl-λ^3-phosphorins react to form π-complexes. In the case of 2.4.6-triphenyl-λ^3-phosphorin, the π-complex *56* is a crystalline, deep red compound of m.p. 156–158 °C (dec.) [58)]. The elemental analysis and spectral properties of *56* are consistent with the proposed π-structure. In the ^{1}H-NMR spectrum the shift of the C−3 and C−5

[structure 55 with Cr(CO)$_5$]   [structure 56 with —Cr(CO)$_3$]

55   *56*

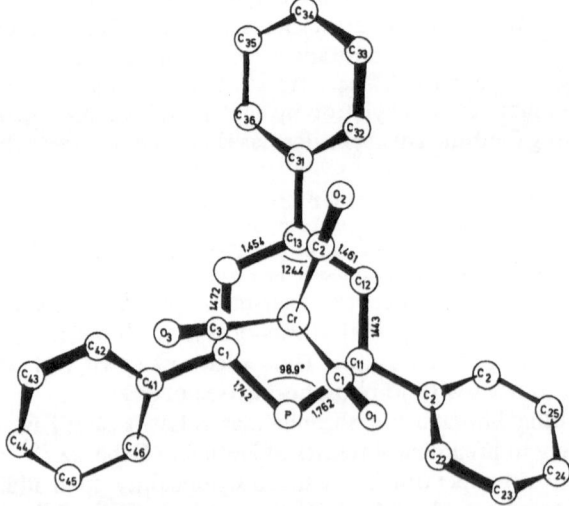

Fig. 16. Crystal structure of 2.4.6-triphenyl-chromium-tricarbonyl-λ^3-phosphorin [59)]

protons from $\delta = 8,07$ to $\delta = 6,00$ ppm, and the concomitant lowering of the J_{P-H} value from 6 to 4,5 Hz provide strong evidence for the formation of a π-complex. Such complexation is also evidenced by the ^{31}P resonance shift from $\delta = -178,2$ to $\delta = +4,3$ ppm.

The crystal structure determination by Vahrenkamp and Nöth [59] proves the assumed structure of this new and interesting chromium-tricarbonyl complex 56 (Fig. 16).

2.4.6-Triphenyl-λ^3-phosphorin interacts with iodine or other polarizable electron donors, as well as with such electron acceptors as tetrachloro-p-benzoquinone and tetracyanoethylene, to produce deeply colored solutions. Such coloration points to the formation of charge-transfer complexes (see p. 43). In some cases electron transfer occurs with the formation of 2.4.6-triphenyl-λ^3-phosphorin cation radical and tetracyanoethylene anion radical. Weber [63] is currently investigating the details of these reactions (see p. 43).

2. Electron Transfer Reactions

a) Radical Cations

The position of phosphorus with respect to nitrogen in the periodic table led to the expectation that it should be much easier to remove one electron from λ^3-phosphorins than from pyridines (see also p. 37). Indeed, soon after the synthesis of 2.4.6-triphenyl-λ^3-phosphorin 20 by Märkl, we discovered that addition of 2.4.6-triphenoxyl 57 [64] in benzene induces oxidation to the very stable radical cation 58 [60]:

20	57	58	59

This transformation can be monitored by ESR-spectroscopy (Figs. 17 and 18). The intense singlet due to 57 disappears, and a new doublet of equally intense signals with a coupling constant of 23.2 Gauss and a complex hyperfine structure appears. The large coupling constant is due to the interaction of the lone electron with phosphorus; the hyperfine structure arises from coupling with the 17 protons.

41

The novel delocalized λ^3-phosphorin radical cation *58* is unusually stable and quite insensitive to air.

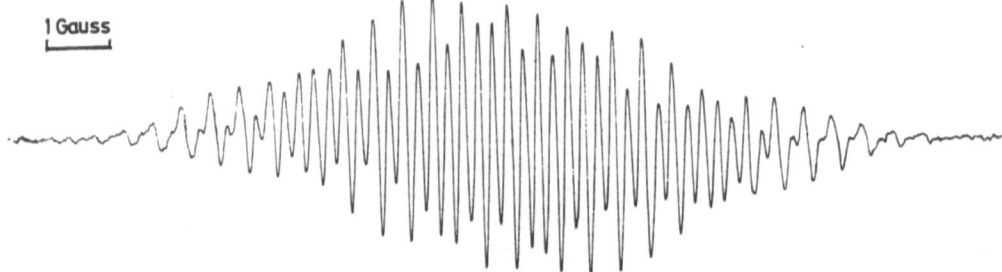

Fig. 17. ESR spectrum of 2.4.6-triphenylphenoxyl

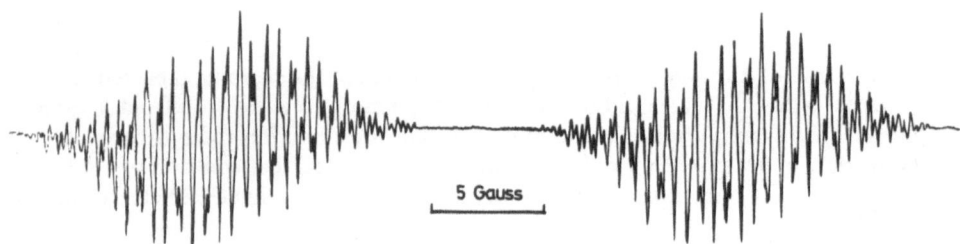

Fig. 18. ESR spectrum of 2.4.6-triphenyl-λ^3-phosphorin radical cation

That this oxidation leads to the radical cation *58* and not to a λ^4-phosphorin-oxide radical *59* was evidenced by the observation that ^{17}O labelled triphenyl-phenoxy radicals do not lead to any ^{17}O coupling pattern, but rather to precisely the same ESR spectrum as before. Nevertheless, the possibility of an equilibrium with the neutral radical *59* cannot be excluded with certainty since the ^{17}O coupling may be quite small and the P-coupling may not be changed very much.

Later on we could show that numerous other *oxidizing agents* can be employed to produce *58*; in all cases identical ESR spectra were obtained. Indeed, bubbling air through a solution of 2.4.6-triphenyl-λ^3-phosphorin in benzene or acetonitrile for some time results in formation of the radical cation *58*. Useful oxidizing agents are $AuCl_3$ or $Hg(OAc)_2$ in DMF, lead tetrabenzoate in benzene, or a suspension of lead oxide in benzene in the presence of a small amount of 2.4.6-triphenylphenol as a redox catalyst. With Hg(II) or Pb(IV) salts, the oxidation is somewhat slug-gish; several hours are required to produce higher concentrations of the radical ca-tion.

Städe [45, 61)] observed an interesting oxidation with *tetrachloro-p-benzoquinone*. In methylene chloride an intense red coloration appears, but no signal in the ESR spectrum. Apparently only a charge-transfer complex *61* is formed, without electron transfer. A similar observation has been made in the reaction of N, N, N', N'-tetramethyl-p-phenylenediamine with tetrachloro-p-benzoquinone in non-polar solvents [62)]. Here, as in our case, electron transfer does not take place until a polar solvent such as acetonitrile is added. The ESR spectrum initially shows the doublet of *58* (23,2 Gauss) overlapping with the sharp singlet of tetrachloro-semiquinone *62* (which has a somewhat smaller g factor). The semiquinone signal slowly disappears until finally only the doublet of *58* remains. The following scheme summarizes the reaction course:

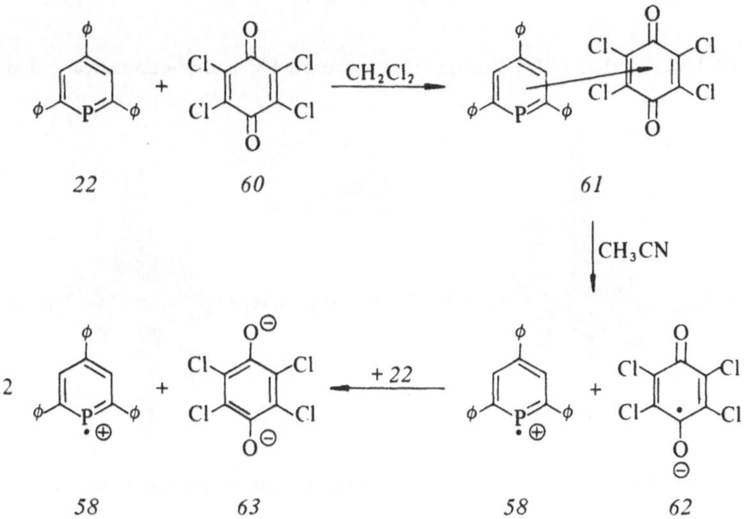

According to Weber [63)] the course of the oxidation of 2.4.6-triphenyl-λ^3-phosphorin with tetracyanoethylene is similar.

Despite its complexity, the ESR spectrum (Fig. 18) of the 2.4.6-triphenyl-λ^3-phosphorin radical cation *58* could be analyzed in full detail. In principle we proceeded in the same way as in the previous case of 2.4.6-triphenylphenoxy radical [64, 45)]: Replacement of the three phenyl substituents by penta-deuterophenyl groups leads to the ESR spectrum in Fig. 19. The coupling constant due to the protons in positions 3 and 5 can easily be determined by inspection of the twofold triplet. The spectrum is quite similar to that of the 2.4.6-tri-tert-butyl-λ^3-phosphorin radical cation (Fig. 20), in which the hyperfine structure of the tert-butyl groups (a_p = 0,8 Gauss) can be detected. Here, a_p = 26,9 Gauss and a_H (protons in 3 and 5 positions) = 2,9 Gauss.

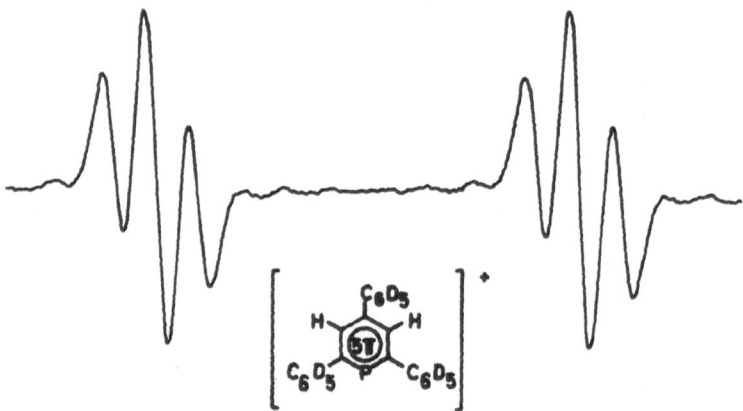

Fig. 19. ESR spectrum of 2.4.6-tris-pentadeuterophenyl-λ^3-phosphorin radical cation

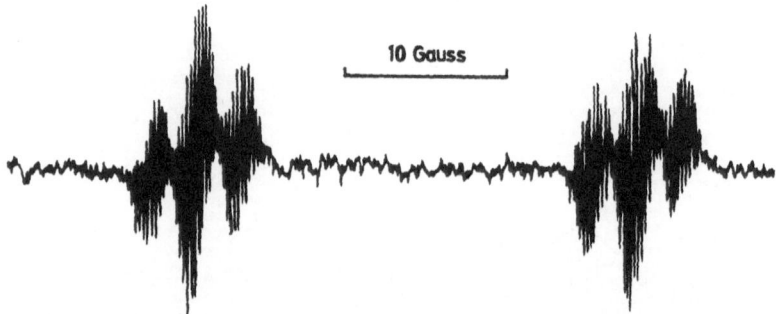

Fig. 20. ESR spectrum of 2.4.6-tri-tert-butyl-λ^3-phosphorin radical cation

By systematically replacing the phenyl groups in positions 2,4 and 6 by pentadeuterophenyl groups and the H of the smallest coupling constant by tert.-butyl groups (*e. g. 64*), the coupling constants of all hydrogen atoms could be determined.

64

64. It is interesting to compare the coupling constants of the 2.4.6-triphenyl-λ^3-phosphorin radical cation *58* with those of the 2.4.6-triphenylphenoxy radical *57* [64] and the 2.4.6-triphenylpyryl radical *32* [31]:

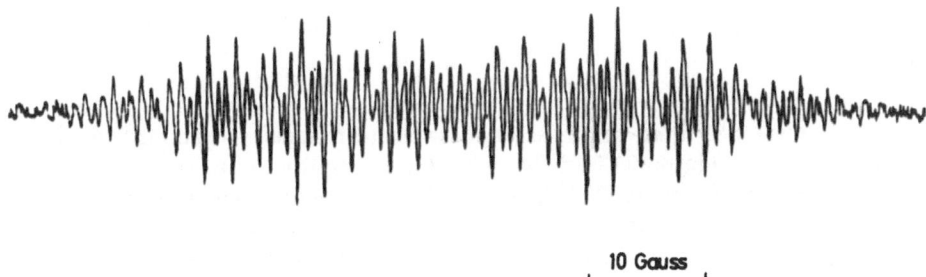

By using the McConnell equation [65] and the experimental ^{1}H-coupling constants, it is possible to calculate the spin density at the carbon atoms bearing protons. The missing parameter which is needed to calculate the spin density at the P atom has been estimated by Thomson and Kilcast [66].

Other 2.4.6-trisubstituted λ^3-phosphorins can also be oxidized to stable radical cations. Examples are 2.4.6-tri-tert-butyl-λ^3-phosphorin *24* (Fig. 20) [44] and 2.6-dimethyl-4-phenyl-λ^3-phosphorin *50* (Fig. 21) [67]. Whereas the ESR spectrum of the radical cation of *24* is easy to interpret (see p. 43), that of *50* is rather complex. Using a spectrum expansion of 81 Gauss, the coupling constant was determined to be a_p = 23,2 Gauss. A septet due to the six equivalent methyl protons with a_H = 7,5 Gauss and intensities 1:6:15:20:15:6:1, as well as an overlapping sextet with a_H = 2,2 Gauss can also be observed. In analogy to other λ^3-phosphorin radical cations, we ascribe these signals to the two 3,5 protons of the central ring and the three ortho and para protons of the phenyl ring, which happen to be degenerate. Finally, the splitting of the triplet due to the two meta protons of the 4-phenyl ring has a value of 0,9 Gauss.

10 Gauss

Fig. 21. ESR spectrum of 2.6-dimethyl-4-phenyl-λ^3-phosphorin radical cation

b) Radical Anions, Dianions, and Radical Trianions

Whereas the oxidation of λ^3-phosphorins to radical cations requires no special precautionary measures, the reduction to radical anions must be carefully performed, particularly as regards the complete exclusion of oxygen and moisture. We operate in a sealed apparatus without any stopcocks; all trace of oxygen have to be removed by repeatedly applying a freeze-thaw procedure under vacuum and by distilling the solvent. If a THF solution of 2.4.6-triphenyl-λ^3-phosphorin is then allowed to flow over a K/Na mirror, one can observe after a short time a growing doublet (32,4 Gauss) which can be ascribed to the coupling of the lone electron with the phosphorus [68]. The ^{1}H hyperfine structure is usually not very well defined. A degree of resolution which would enable all coupling constants to be determined has not yet been achieved.

If the above-mentioned radical solution is allowed to come in contact with the K/Na mirror once again, then a point is reached where no ESR signal can be observed. Further contact with the mirror results in the appearance of a new signal (Fig. 22).

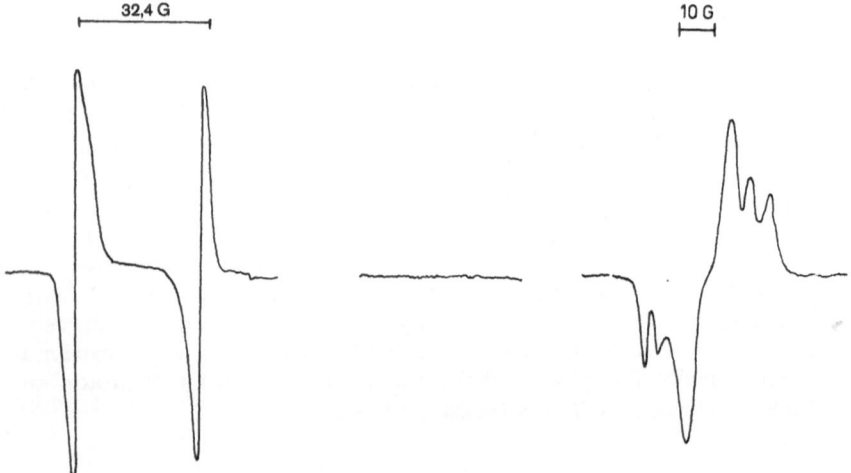

Fig. 22. ESR spectra of the radical anion, dianion and radical trianion of 2.4.6-triphenyl-λ^3-phosphorin

We interpret [69] these results as stepwise electron transfer reactions. 2.4.6-Triphenyl-λ^3-phosphorin 22 initially picks up one electron to form the radical anion 65, which with more K/Na is then reduced to the diamagnetic anion 66. A third electron is then transferred from the contact with the K–Na alloy, forming the paramagnetic radical trianion 67. The phosphorus coupling constant in 67 of $a_p = 4,6$ was calculated from the ESR spectrum of 2.4.6-tris-pentadeutero-phenyl-λ^3-phosphorin.

$$\overset{58}{\underset{(a_p = 23.2G)}{5\pi \oplus}} \xleftarrow{-e} \overset{22}{6\pi} \xrightarrow{+e} \overset{65}{\underset{(a_p = 32.4G)}{7\pi \ominus}} \xrightarrow{+e} \overset{66}{8\pi^{2\ominus}} \xrightarrow{+e} \overset{67}{\underset{(a_p = 4.6G)}{9\pi^{3\ominus}}}$$

Our interpretation is strengthened by the following observations:

1) Upon mixing equivalent amounts of separately prepared solutions of the radical cation 58 (a_p = 23,2 Gauss) and the radical anion 65 (a_p = 32,4 Gauss), the ESR signal disappears. The typical absorption of 2.4.6-triphenyl-λ^3-phosphorin 22 can be seen in the UV spectrum.

2) If a solution of the radical trianion 67 (prepared by the K/Na method) is "titrated" with a solution of the λ^3-phosphorin 22, then upon addition of one aliquot of 22 to 2 aliquots of 67 no ESR signal can be seen (formation of the dianion 66). Continuation of the titration results in the appearance of an ESR signal due to the radical anion 65 (a_p = 32,4 Gauss), which reaches maximum intensity after addition of one equivalent of 22.

Of course, in all of these experiments sealed, high-vacuum glass systems with complete exclusion of air and moisture have to be employed. Under such conditions the results are easily reproducible; the radical ions themselves are indeed very stable. The radical monoanion 65 is best prepared by first reducing all the way to the radical trianion 67, and then adding the double amount of starting solution of 2.4.6-triphenyl-λ^3-phosphorin.

2.4.6-triphenyl-λ^3-phosphorin can also be directly reduced with sodium in THF. In this case the reduction stops at the level of the radical monoanion [63].

Table 7. [31]P coupling constants of some λ^3-phosphorin radical cations and anions

R^2	R^4	R^6	a_P (Gauss)	
			Radikal-cation	Radikal-anion
C_6H_5	C_6H_5	C_6H_5	23.2	32.4
C_6D_5	C_6H_5	C_6D_5	23.2	32.4
C_6D_5	C_6D_5	C_6D_5	23.2	32.4
C_6H_5	$C(CH_3)_3$	C_6H_5	23.7	31.8
$C(CH_3)_3$	C_6H_5	$C(CH_3)_3$	23.7	30.5
$C(CH_3)_3$	$C_6H_4OCH_3(4)$	$C(CH_3)_3$	23.3	29.1
$C(CH_3)_3$	$C(CH_3)_3$	$C(CH_3)_3$	26.9	27.8

2.4.6-Tri-tert-butyl-λ^3-phosphorin *24* in THF could not be reduced to a radical trianion with K/Na. Only the doublet of the radical monoanion can be seen in the ESR spectrum (a_p = 27,8 Gauss). Good resolution of the hyperfine structure of the signals was not obtained.

So far chemical reactions have been performed only with the cation radicals as intermediates.

3. Oxidation with Oxygen, Nitric Acids, Hydrogen Peroxide, Halogens and Aryldiazonium Salts

a) Oxidation with Oxygen

α) *Air Oxidation in the Dark.* In the solid state 2.4.6-triphenyl or higher substituted λ^3-phosphorins are rather stable to air. However, if a solution of 2.4.6-triphenyl-λ^3-phosphorin *22* in benzene (which dissolves relatively large amounts of oxygen is allowed to stand for about one week at room temperature in the dark, two crystalline oxidation products of m.p. 162–165 °C and 225 °C precipitate, as Vogel has found [70]. The constitution of the two compounds was published by us with some reservations. When Hettche [88b] recently studied this reaction again, most of the experimental data were confirmed. The constitution of the higher melting compound is still in question, but the structure of the lower melting substance is now fully confirmed as the 4.4'-peroxy-bis-phosphinic acid *68*.

Careful elementary analysis, especially for oxygen, is in accord with the formula. UV (λ_{max} 245 nm, ϵ = 3600) proves 4-substituted phosphacyclohexadiene (2,5) structure. Thus, ^{1}H-NMR shows a doublet for the four equivalent protons at C–3, 3'. 5, 5' at δ = 6,8 ppm, J_{P-H} = 35 Hz. IR secures P–OH (2280/cm) and P = O (1170/cm) groups.

Alkylation of *68* with triethyloxonium tetrafluorborate leads to three stereoisomeric 4.4'-peroxy-phosphinic acid ethyl esters *69a–c*. One (*69a*) could be isolated as a pure substance, m.p. 178–180 °C; the existence of the two others, m.p. 145–152 °C, is proved by ^{1}H–NMR, although they could not be fully separated.

The three isomers are *cis/trans* isomers with respect to the substituents -phenyl and -O—O at C—4, and -OC$_2$H$_5$, = O at the P atom. For convenience, we use the E/Z nomenclature recommended for ethylene diastereoisomer assignment. If we replace the plane in ethylenes through the sp^2 and sp^3 atoms by a plane vertical to the phosphacyclohexadiene (2,4) ring (thought of in first approximation as planar) through C—4 and the phosphorus atom, we have to denote the isomer with -O—O at C—4 and = O at P on the same side of this plane as "Z", and that with -O—O at C—4 and -OC$_2$H$_5$ at P on different sides as "E" diasteroisomer.

69 therefore should give three stereoisomers E—E, E—Z and Z—Z, all of which we have found. In *68*, which loses its stereoisomerism by epimerization of the P substituents through hydrogen bridges of the phosphinic acid, the three steric diastereoisomers cannot de detected.

For assignment, we apply the same arguments as are used on for the diastereo-isomeric 4-hydroxy-phosphinic-esters *72 Z* and *72 E*, *i. e.* ^{1}H-NMR shifts of the protons at C—3 and C—5 and the protons of the CH$_3$ groups of the P—O—CH$_2$—CH$_3$ group (Table 8):

Table 8. ^{1}H-NMR shifts of peroxy-ethylester *69* diastereomers

		H at C—3 and C—5 (4H, d)		H of CH$_3$(O—CH$_2$—CH$_3$) (6H, t)	
		(ppm)	J_{P-H}	(ppm)	J_{H-H}
69a	E—E	6.81	35	0.95	7
69b	E—Z	6.87	35	0,97	7
		6.88	35	0.82	7
69c	Z—Z	6.96	35	0.76	7

Reduction of a mixture of *69a—c* or *69a* with zinc dust in acetic acid yields two diastereoisomeric 4-hydroxy-phosphinic acid ethyl esters *70 Z* and E, or only *70* E.

70 Z   *70 E*   *71*

The same compounds were prepared by Hettche from 1.1-diethoxy-2.4.6-triphenyl-λ^5-phosphorin, using the method of Städe (see p. 124). The structures *70 Z* and *70 E* are fully in accord with all analytical and spectroscopic data, especially the UV absorption at short waves, and in the ^{1}H-NMR spectra the equivalence of the protons at C—3 and C—5; both sets of data exclude the structure of 2-hydroxy-phosphinic ester.

The stereomerism was studied in detail by Städe with the methylesters *72* and *70*.

1) With trifluoroacetic acid the two stereoisomers *72* Z and *72* E give the same deep blue cation *73*. Adding water then gives a *mixture* of the two diastereo-isomers (epimerization at C—4). With alcohols the cation affords a diastereome-

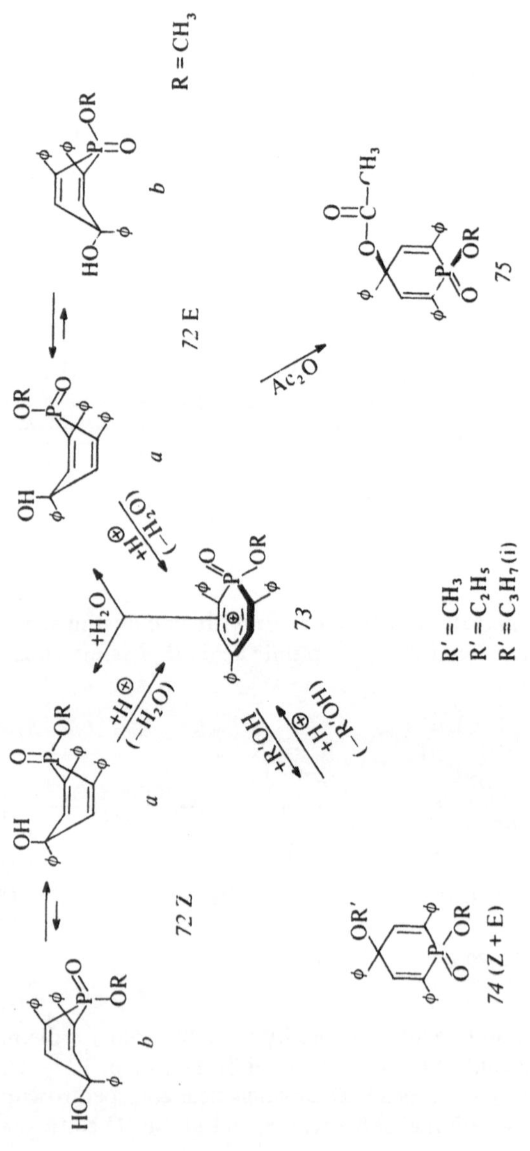

ric mixture of 4-alkoxy-phosphinic esters *74a–c* Z and E. By cautious acylation of the stereoisomer *72* E with acetanhydride in pyridine, the acetoxy compound *75* could be prepared:

The formation of blue cations like *73* is a very general type of reaction. It can be used as a characteristic reaction with all 2.4.6-aryl-substituted λ^5-phosphacyclohexadiene compounds with an OH or OR group at position 2 or 4. An analogous reaction is well known from experiments with 2.4.6-triphenyl-4-hydroxy- (or alkoxy) cyclohexadiene(2,5)-one *76* which also gives a deep blue cation *77* with strong acids. With water or alcohols it can be reconverted to the hydroxy or alkoxy compounds *76*[121].

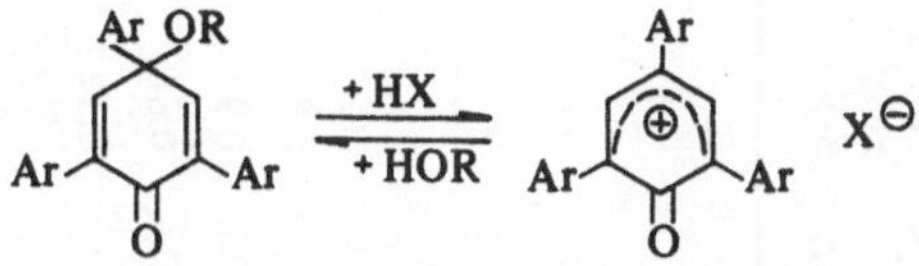

76 *77*

a) R = H
b) R = CH$_3$
c) R = C$_2$H$_5$
d) R = CH(CH$_3$)$_2$

2) The steric assignment of the isolated compounds of type *70*, *72* of *74* to the Z or E configuration is difficult, because the exact conformation of the phosphacyclohexadiene ring is not known. The Z as well as the E diastereomer can exist in two conformations, a and b, in solution. According to ^{1}H-NMR studies, we conclude that *only one conformer is preferred*, i. e. *72 Z a* and *72 E a*.

We presume that *the phenyl substituents* at C–2, 4 and 6 are *in an equatorial position*[71,72]. Then, by ^{1}H-NMR, a reasonable assignment is possible. When we look at the chemical shift of the protons of the CH$_3$ of the P–OCH$_3$, or more exactly of the CH$_3$ of P–O–CH$_2$–CH$_3$, we find that only in the Z position (*72 Z a*) these protons lie over the aromatic rings in 2 or 6 position. The NMR signals therefore are shifted to higher field. At the same time the P=O group will shift the protons at C–3 and C–5 to lower field. This would be in accord with the low field shift of the ethytene protons in bicycloheptadiene derivatives *78* and *79*

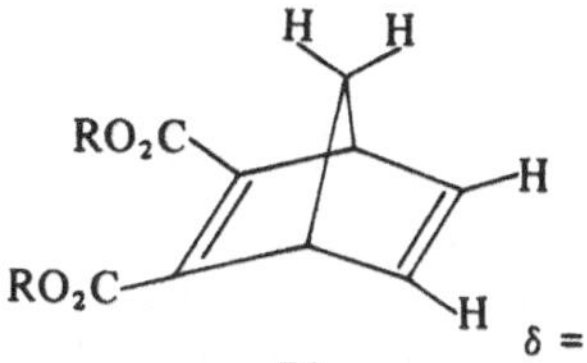

78 δ = 6.1 ppm *79* δ = 7.6 ppm

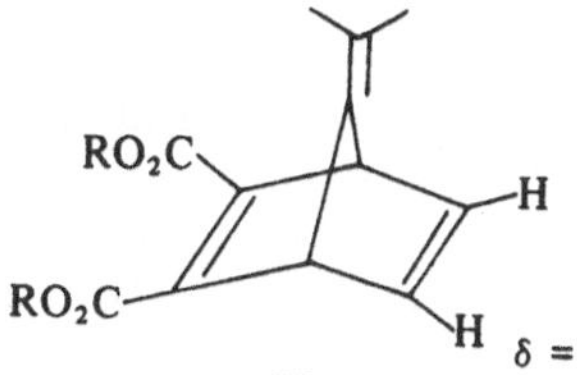

51

Table 9. ¹H-NMR shifts of Z and E diastereomers type 72 δ in ppm

R^1	$R^2 = R^4 = R^6$	$R^{4'}$	m. p. (°C)	Proton at C-3 and C-5	Proton at R^1 α-pos.	Proton at R^1 β-pos.	Stereo chem- proposed	IR (cm⁻¹) (3584) sharp	(3256) broad
CH_3	C_6H_5	H	198–200	6.35	3.16	–	Z	+	–
CH_3	C_6H_5	H	194–195	6.31	3.24	–	E	+	‡
CH_3	$C_6H_4CH_3$	H	218–223	6.67	3.26	–	Z	+	–
CH_3	$C_6H_4CH_3$	H	202–206	6.63	3.39	–	E	+	‡
CH_3	$C_6H_4OCH_3$ [b]	H	177–179	6.64	3.38	–	Z		
CH_3	$C_6H_4OCH_3$ [c]	H	179–181	6.60	3.68	–	E		
C_2H_5	C_6H_5	H	239–241	6.76	3.72	0.8	Z		
C_2H_5	C_6H_5	H	194–196	6.72	3.8	0.94	E		
$C_3H_7(i)$	C_6H_5	H	221–228	6.77	–(m)	0.64	Z	+	–
$C_3H_7(i)$	C_6H_5	H	204	6.64	–(m)	0.92	E	3572 / +	3256 / ‡
CH_3	$C(CH_3)_3$	H	228	6.69	3.56	–	Z	–	–
CH_3	$C(CH_3)_3$	H	274	6.57	3.74	–	E	–	–
C_2H_5	$C(CH_3)_3$ [d]	H	180	6.69	3.74	1.4	Z	–	–
C_2H_5	$C(CH_3)_3$	H	190	6.55	4.05	1.4	E	–	–

Indeed, in all diastereoisomers of type 72 we established an *inverse* shift of the ring protons at C–3, C–5, and the protons of the P–O–R group (Table 9). The assignment "Z" or "E" is justified only when the "a" conformations in 72 Z and 72 E are the favoured ones. On the other hand, axial CO–CH₃ groups in cyclohexane series come at lower field than equatorial H of CO–CH₃ which fits to the lowfield absorption of the E-ester CH₃ group.

Table 9 (continued)

R¹	R² = R⁴ = R⁶	R⁴'	m. p. (°C)	Proton at C–3 and C–5	Proton at R¹ α-pos.	H at R⁴' α-	H at R⁴' β-	IR (cm⁻¹) (3584) sharp	IR (cm⁻¹) (3256) broad
CH_3	C_6H_5	CH_3	98–101.5	6.78	3.40	3.58	–	–	–
CH_3	C_6H_5	CH_3		6.71	3.38	3.38	–	–	–
CH_3	C_6H_5	C_2H_5	108–111	6.19	3.18	3.50	1.19	–	–
CH_3	C_6H_5	C_2H_5		6.13	3.22	3.54	1.31	–	–
CH_3	C_6H_5	$C_3H_7(i)$	Oil	6.61	3.38	m	1.23	–	–
CH_3	C_6H_5	$C_3H_7(i)$		6.59	3.46	m	1.39	–	–
CH_3[a]	C_6H_5	$COCH_3$	140–142	6.72	3.43	–	2.27	–	–

a) From 72 E.
b) C_6H_5 in pos. 4.
c) $C_6H_4OCH_3$ in pos. 4.

Table 9 also contains the ¹H-NMR data for some derivatives 74 and 75, for which now definite stereochemical assignment is possible.

The IR spectra of each diastereoisomer pair also show characteristic differences, especially in the O–H region. Whereas the Z compounds (Table 9) have only a sharp band near 3590/cm in CCl_4, the E compounds have only a weak band near 3590/cm but an intense broad absorption at 3270/cm. The P = O absorption at 1225/cm is also slightly shifted to lower frequencies. On dilution Städe[45] found that the intensity of the broad long-wave band is diminished and at the same time the short-wave band strengthened. No doubt, the observed long-wave absorption comes from an intermolecular association over hydrogen bridges of the E diastereomers.

It is possible that the higher-melting Z diastereomers also have an intramolecular hydrogen bridge.

Turning back to the stereochemistry of the three diastereoisomeric peroxy esters *69*, we now can assign stereochemistry by the ^{1}H-NMR shifts, as was done in Table 8, p. 49, in accordance with all other observations.

β) *Oxidation with Singlet Oxygen (Light and Sensitizer).* 2.4.6-Triphenyl-λ^3-phosphorin *22*, when oxygenated in benzene or in hexane/methanol in the presence of eosin, methylene blue or rose bengal, gave a rather complex mixture of different oxidation products. 2.4.6-Tri-tert-butyl-λ^3-phosphorin *24*, however, yields, as Chatzidakis and Schaffer have found[75], two crystalline oxidation products, m.p. 167 °C and 152 °C, respectively. They can be isolated in about 15% each. The higher-melting compound proved to be identical with 4-hydroxy-phosphinic acid *81a*, previously described by Mach[44] p. 55). In cyclohexane/methanol the methylester *81b*, also described by, Mach is obtained.

The analytical and spectroscopic data of the lower-melting compound are in accord with structure *82*. We therefore suppose that in the first step addition of the singlet oxygen takes place at the C, position 4, and at the phosphorus of *24*, leading to the endoperoxide *80*. This intermediate, which we could not isolate, is then converted by hydrolysis or alcoholysis to *81a* or *81b*, or it rearranges to the endoxy-λ^5-phosphinoxide *82*. Analogous processes with carbon compounds have been discussed by Schuler-Elte and others [76]. The course of this photo-oxidation is noteworthy since it represents the first example of 1.4 addition in which a heteroatom is involved.

b) Oxidation with Nitric Acid

As Mach [44] has found, oxidation of 2.4.6-tri-tert-butyl-λ^3-phosphorin *24* in glacial acetic acid with a mixture of equal parts of conc. nitric and sulfuric acids yields the 4-hydroxy-phosphinic acid *83*. A small amount (2%) of the 2-hydro-phos-phinic acid of type *90b* (see p. 60) is also formed. The acid *83* (m.p. 167° C) crystallizes with one mole H_2O.

$$24 \xrightarrow{HNO_3/H_2SO_4} 83 \xrightarrow{CH_2N_2} 84\ E \qquad 84\ Z$$

The structure of the acid *83* is supported by elemental analysis. ^{1}H-NMR shows only one doublet for the two (equivalent) protons at C$-$3 and C$-$5 (δ = 6,58 ppm, J_{P-H} = 36 Hz) and two different singlets (δ = 1,12 and δ = 1,44 ppm, 1:2) for the three tert. butyl groups. ^{31}P: δ = $-16,35$ ppm (in pyridine with H_3PO_4 as external standard).

By esterification with diazomethane two stereoisomeric methylesters *84* E and *84* Z were found. They could be separated by thin-layer chromatography: m.p. 279 °C and 278 °C (sealed tube). The UV spectra are nearly the same, but the ^{1}H-NMR spectra differ considerably (Table 10).

Table 10. ^{1}H-NMR spectra of stereoisomeric 4-hydroxy-2.4.6-tri-tert-butyl-phosphinic acid methylesters *84* E and *84* Z (δ in ppm, CDCl$_3$ solvent)

	H at C$-$3 and C$-$5 (d, 2H)	OCH$_3$ (d, 3H)	C$-$OH (s, 1H)	Tert-butyl- group at C$-$2 and C$-$6 (s, 18H)	Tert-butyl- group at C$-$4 (s, [a], 9H)
84 E	6.57 (*J* = 37Hz)	3.74 (*J* = 12Hz)	1.88	1.37	1.05
84 Z	6.69 (*J* = 37Hz)	3.56 (*J* = 11Hz)	2.03	1.41	1.13

[a] Small coupling with ^{31}P.

If we again assume equatorial positions for the large substituents in 2.4.6. position, then *84* E and *84* Z should represent the stable conformations of the diastereomers. Applying the same arguments as in the phenyl series *72* for the Z configuration of *84* at C$-$3,5, the proton signals should be found at lower field. That the protons of the P$-$O$-$CH$_3$ group are shifted to higher field is due to the high shielding of the 2,6-tert.-butyl groups (in the phenyl series by the anisotropic effect caused by 2,6-phenyl groups). In tert.-butyl series this effect gets smaller with β-CH$_3$ protons; in the phenyl series it gets larger.

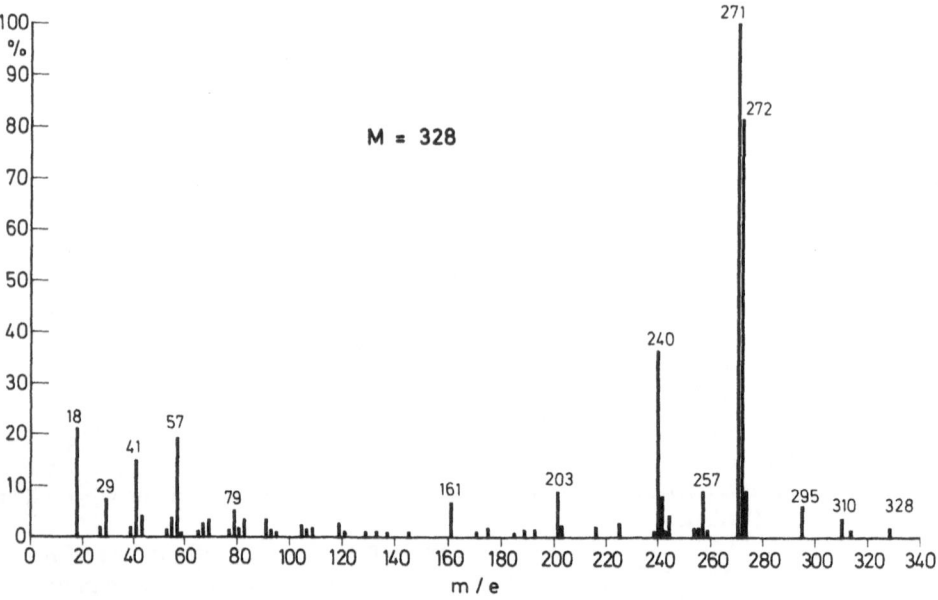

Fig. 23. Mass spectrum of E-2.4.6-tri-tert-butyl-phosphinic acid methyl ester

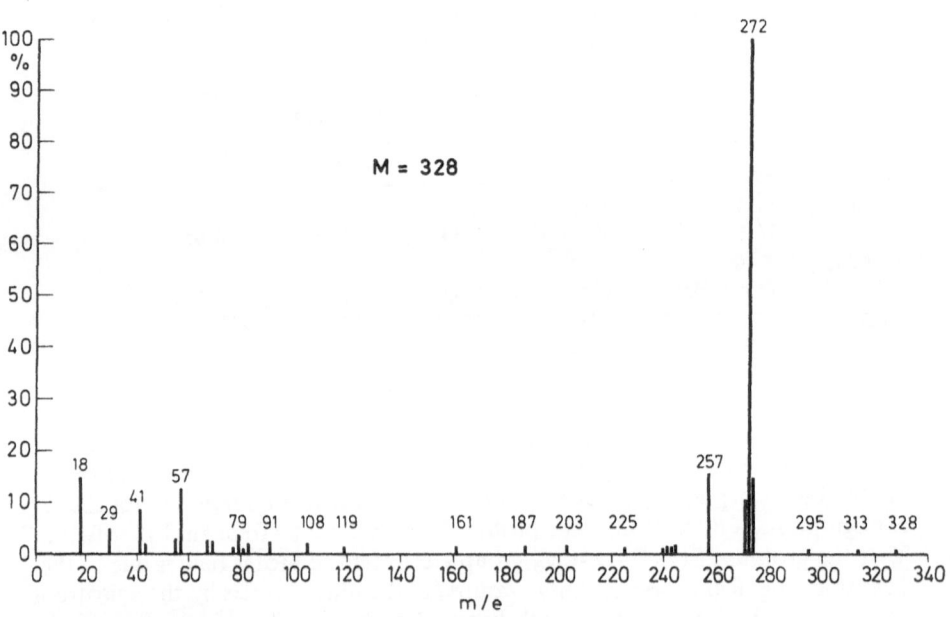

Fig. 24. Mass spectrum of Z-2.4.6-tri-tert-butyl-λ⁵-phosphinic acid methyl ester

In the tert.-butyl series no significant difference in the IR spectra of the two diastereoisomers can be detected. This is in a sharp contrast to the corresponding aryl series, *i. e.* *72* E and *72* Z, and must be influenced by the large hydrophobic tert.-butyl groups which prevent association by OH bridges.

On the other hand the mass spectra of the two isomers differ very significantly only in the tert.-butyl series (Figs. 23 and 24).

The compound having E configuration splits off H_2O (328 → 310) by hydrogen transfer from the CH_3–O- to the OH at the same side of the ring: the Z configuration does not. The E configuration not only loses isobutylene (56: 328 → 272) by hydrogen transfer from any one of the tert-butyl groups in 2,6 (or 4) to oxygen on phosphorus, but also a tert-butyl *radical* (57: 328–271) forming the stable carbonium-phosphonium-oxonium ion, resembling the cation *73* in the phenyl series. The loss of CH_3O of the E configuration to the radical cation m/e = 240 is another characteristic feature not observed in the Z configuration.

As far as we know, differences in mass spectroscopy of stereoisomers, known first from Bieman's study of exo- and endoborneol, have only been expressed in the relative intensities of the peaks, not in a different pattern [73].

The unique situation in the phosphorus compounds which we also observed with other esters of type *84* E and *84* Z (C_2H_5 instead of CH_3) may be caused (more favored in the tert-butyl series) by a much better localization of the electron of the radical ion at the phosphorus than in carbon series with delocalized π bonds.

Table 11 contains the known 4-hydroxy-phosphinic acids and esters of type *72* and *84*, prepared by different methods.

Table 11. 4-Hydroxy-phosphinic acids *72* and derivatives

R^1	R^2	R^4	R$^{4'}$	R^6	m. p. °C	Lit.	Method
H	C(CH$_3$)$_3$	C(CH$_3$)$_3$	OH	C(CH$_3$)$_3$	167	44	B
CH$_3$	C(CH$_3$)$_3$	C(CH$_3$)$_3$	OH	C(CH$_3$)$_3$	Z 288 [a] E 274 [a]	44	B
C$_2$H$_5$	C(CH$_3$)$_3$	C(CH$_3$)$_3$	OH	C(CH$_3$)$_3$	Z 180 [a] E 190 [a]	44	C, D
CH$_3$	C(CH$_3$)$_3$	C$_6$H$_4$OCH$_3$	OH	C(CH$_3$)$_3$	225	44	B
C$_6$H$_5$	C(CH$_3$)$_3$	C$_6$H$_4$OCH$_3$	OH	C(CH$_3$)$_3$	197	44	E
H	C$_6$H$_5$	C$_6$H$_5$	OH	C$_6$H$_5$	–	88b	E
CH$_3$	C$_6$H$_5$	C$_6$H$_5$	OH	C$_6$H$_5$	Z 198 E 194–95	45	A

Table 11 (continued)

R^1	R^2	R^4	R$^{4'}$	R^6	m. p. °C	Lit.	Method
CH$_2$CH$_3$	C$_6$H$_5$	C$_6$H$_5$	OH	C$_6$H$_5$	Z 239–41 E 194–96	88b	A, E
CH$_2$-CH$_2$OH	C$_6$H$_5$	C$_6$H$_5$	OH	C$_6$H$_5$	205	45	A
CD$_2$CD$_2$OH	C$_6$H$_5$	C$_6$H$_5$	OH	C$_6$H$_5$	207	45	A
C(CH$_3$)$_2$OH	C$_6$H$_5$	C$_6$H$_5$	OH	C$_6$H$_5$	179–80	45	A
CH$_2$-CH$_2$-CH$_2$Br	C$_6$H$_5$	C$_6$H$_5$	OH	C$_6$H$_5$	Z 209–13 E 172–72	45	A
CH$_2$-CH$_2$-CH$_2$OH	C$_6$H$_5$	C$_6$H$_5$	OH	C$_6$H$_5$	Z 163–65 E 150–52	45	A
CH(CH$_3$)$_2$	C$_6$H$_5$	C$_6$H$_5$	OH	C$_6$H$_5$	Z 221– 28 E 204	45	A
CH$_3$	C$_6$H$_4$OCH$_3$	C$_6$H$_5$	OH	C$_6$H$_4$OCH$_3$	Z 177–79 E 179–81	45	A
CH$_3$	C$_6$H$_4$CH$_3$	C$_6$H$_4$CH$_3$	OH	C$_6$H$_4$CH$_3$	Z 218–23 E 202– 06	45	A
CH$_3$	C$_6$H$_5$	C$_6$H$_5$	OCH$_3$	C$_6$H$_5$	98–101.5 (2 isomers)	45	A
CH$_3$	C$_6$H$_5$	C$_6$H$_5$	OC$_2$H$_5$	C$_6$H$_5$	108–11 (2 isomers)	45	A
CH$_3$	C$_6$H$_5$	C$_6$H$_5$	OCH(CH)$_3$	C$_6$H$_5$	oil (2 isomers)	45	A
CH$_3$	C$_6$H$_5$	C$_6$H$_5$	OCOCH$_3$	C$_6$H$_5$	140– 42	45	A
C$_2$H$_5$	C(CH$_3$)$_3$	C(CH$_3$)$_3$	Cl	C(CH$_3$)$_3$	155	44	D
OR^1 = Cl	C(CH$_3$)$_3$	C(CH$_3$)$_3$	Cl	C(CH$_3$)$_3$	77–78.5	44	D
OR^1 = C$_6$H$_5$	C$_6$H$_5$	C$_6$H$_5$ OH		C$_6$H$_5$	239–41	86	H
OR^1 = C$_6$H$_5$	C$_6$H$_5$	O		C$_6$H$_5$	142–43	86	H

a) Sublimates, m. p. in closed tube.

Method A from 1,1-dialkoxy-λ^5-phosphorins with LiBr/H$_2$O$_2$/HAc.
Method B from λ^3-phosphorins by HNO$_3$.
Method C from 1.1-dialkoxy-λ^5-phosphorins by HNO$_3$.
Method D from λ^3-phosphorins, chlorination etc.
Method E from λ^3-phosphorins over autoxidation products.
Method H synthetic methods.

c) Oxidation with Hydrogen Peroxide

Mach [44] investigated hydrogen peroxide oxidation of 2.4.6-tri-tert-butyl-λ^3-phos-phorin 24 in analogy to pyridine oxidation to pyridine-N-oxide. Whereas 2.4.6-tri-tert-butyl-pyridine could be recovered unchanged, 2.4.6-tri-tert-butyl-λ^3-phos-phorin under the same conditions is immediately oxidized to the 2-hydro-phos-phinic acid m.p. 203 °C (transformation at 170 °C) 86 (CH$_3$ = H) which, according to mass spectrum and osmometric measurement, dimerizes to *85b*. It seems reasonable to suppose that oxidation leads first to the phosphinoxide-hydrate *85a*.

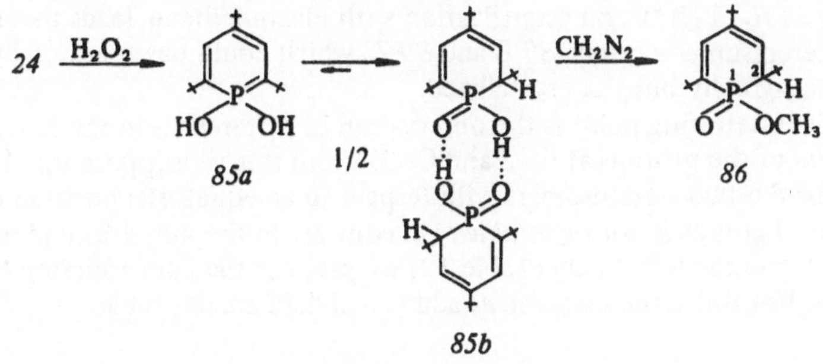

This rearranges by proton shift to the 2-hydro-phosphinic acid *85b*, two molecules of which associate by hydrogen bridging to the dimer. The long-wave maximum at 275 nm ($\epsilon = 4075$), ($\lambda_{max_2} = 237$ nm, $\epsilon = 2925$) confirms the phosphacyclohexadiene (2,4) system.

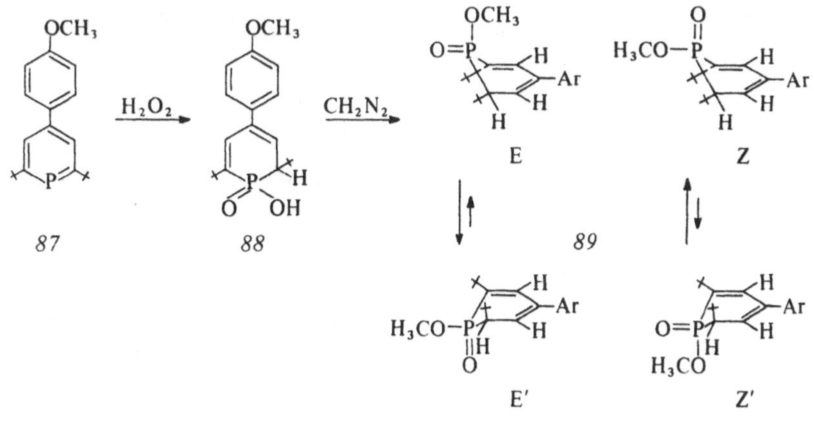

Esterification with diazomethane affords the methyl ester *86* m.p. 51 °C (λ_{max} 280 nm ($\epsilon = 3200$) and 246 nm ($\epsilon = 3200$). No diastereoisomers could be isolated. The ^{1}H-NMR is in accord with structure *86*: three different signals for the three nonequivalent tert-butyl groups at $\delta = 1,29$, $1,10$, and $1,09$ ppm, a doublet at $\delta = 3,54$ ppm ($J_{P-H} = 11$ Hz) for the OCH_3 group and three different signals for the three ring protons: H at C–3 comes at $\delta = 5,75$ ppm as an octet due to P coupling (23 Hz) and H coupling (7 Hz with H at C–2, 2 Hz with H at C–5): H at C–5 appears at $\delta = 6,52$ ppm as a quartet, coupling to phosphorus (36 Hz) and the C–3 proton (2 Hz); H at C–2 at $\delta = 2$ ppm also as a quartet, coupling with phosphorus (23 Hz) and the proton at C–3 (7 Hz). ^{31}P resonance, which again shows the H couplings, comes at δ = 39,8 ppm (in pyridine, H_3PO_4 as external standard).

Analogous results were found in hydrogen peroxide oxidation of 2.6-di-tert-butyl-4(4-methoxyphenyl)-λ^3-phosphorin 87. The crystalline 2-hydrophosphinic acid 88, m.p. 176—178 °C, on esterification with diazomethane, leads to a mixture of two diastereoisomeric esters, 89 E and 89 Z, which could be separated by thin-layer chromatography but not crystallized.

The most interesting point is the observation of differences in the H—H *coupling constants* of the protons at C—2 and C—3. From this we suppose that the *conformation* for the two stereoisomers with respect to an equatorial position of all three tert-butyl groups *is not maintained* (in contrast to the 4-hydroxy-phosphinic acid series). From the NMR data (Table 12) we propose the conformation 89 E' instead 89 E. For full steric assignment additional data are desirable.

Table 12. ^{1}H-NMR of diastereoisomers 89 E and 89 Z (δ in ppm (CDCL$_3$); J in Hz)

	H at C−2	H at C−3	H at C−5	H at OCH$_3$	H at C−5	H at C−2
86 E (?)	2.20; (J_P = 25; J_H = 6	6.04 (J_P = 22; J_H = 6.2)	6.67 (J_P = 36; J_H = 2)	3.63 (J_P = 12)	1.35	1.19
86 Z (?)	2.58; (J_P = 25; J_H = 3)	6.05 (J_P = 20; J_H = 2.3)	6.64 (J_P = 36; J_H = 2)	3.60 (J_P = 17)	1.30	1.20

2.4.6-Triphenyl-λ^3-phosphorin 22, when oxidized with hydrogen peroxide, leads to the noncrystallized 2-hydrophosphinic acid 90b. The tautomers 90a and c are minor components of an equilibrium, as we suppose from the spectroscopic data, especially UV absorption. With strong bases, a highly fluorescent yellow-orange solution arises, indicating the presence of the anion *91*:

Alkylation with triethyl oxonium tetrafluoroborate yields two products (in the proportion 3:1) 1.2-dihydro-1-ethoxy-2-ethyl-2.4.6-triphenyl-phosphorin oxide *92* m.p. 161−162 °C, and 1.1-diethoxy-2.4.6-triphenyl-λ^5-phosphorin *71*, m.p. 106 °C. *71* is known from Städe's synthesis (p. 84), *92* was identified by ^{1}H-NMR. No diastereoisomers were isolated [88b].

Hydrogen peroxide oxidation of other λ^3-phosphorins produces analogous hydrophosphinic acids. Chatzidakis [37], in oxidizing the dihydrophenanthren-λ^3-

Table 13. Derivatives of the 2-hydro-phosphinic acid *90*

R^1	R^2	R^4	R^6	m. p. °C	Method	Lit.
H	$C(CH_3)_3$	$C(CH_3)_3$	$C(CH_3)_3$	203 (dimer)	A	[44]
H	$C(CH_3)_3$	C_6H_5	$C(CH_3)_3$	169	A	[44]
H	$C(CH_3)_3$	$C_6H_4OCH_3$	$C(CH_3)_3$	176−8	A	[44]
H	C_6H_5	C_6H_5	C_6H_5	Oil	A	[88b]
H	$C(CH_3)_3$	C_6H_5	(5) (6)	239−52 mixture with 4H-	A	[37]
CH_3	$C(CH_3)_3$	$C(CH_3)_3$	$C(CH_3)_3$	51	B	[44]
CH_3	$C(CH_3)_3$	C_6H_5	$C(CH_3)_3$	106−7	B	[44]
CH_3	$C(CH_3)_3$	$C_6H_4OCH_3$	$C(CH_3)_3$	Oil (2 isomers)	C, B	[44]
HO	$C(CH_3)_3$	$C_6H_4OCH_3$	$C(CH_3)_3$	180	C	[44]
CH_3	$C(CH_3)_3$	C_6H_4OH	$C(CH_3)_3$	245 decomp.	C	[44]
CH_3	C_6H_5	C_6H_5	C_6H_5	Oil	A	[88]
C_2H_5	C_6H_5	C_6H_5	C_6H_5	Oil	C	[45]
Instead of OR^1 = C_6H_5	C_6H_5	C_6H_5	C_6H_5	156−8	D	[22] [77] [86]

Method A: From λ^3-phosphorins by oxidation with H_2O_2.
Method B: From 2-hydro-phosphinic-*90* with diazomethane.
Method C: From 1.1-dialkoxy-λ^5-phosphorins with BBr_3.
Method D: From pyrylium salts with phenyl phosphine

phosphorin, isolated a hydrophosphinic acid m.p. 239–242 °C, presumed to have the structure of 2-hydro-phosphinic acid.

Koe and Bickelhaupt [33, 74] obtained a 4-hydro-phosphinic acid by hydrogen-peroxide oxidation of 9-chloro-9-λ^5-phospha-9.10-dihydro-anthracene, thus verifying its structure.

Most of the hydrophosphinic acids are unstable in air. They easily autoxidize to hydroxy-phosphinic acids. There are other methods leading to the hydro-phosphinic acids. The known compounds of this type are summarized in Table 13.

d) Oxidation with Halogens

2.4.6-Tri-tert-butyl-λ^3-phosphorin 24 readily reacts with bromine and with chlorine. Mach [44], oxidizing with bromine in CCl_4, could not isolate a crystalline product. The brown colour of the addition product of one mole Br_2 to one mole 24 disappeared with water and the crystalline 2-hydro-phosphinic acid 85b could be isolated in 45% yield. Methyl-magnesium-iodide or red phosphorus yielded 2.4.6-tri-tert-butyl-λ^3-phosphorin 24. It seems reasonable to suppose that on bromination 1.1-dibromo-2.4.6-tri-tert-butyl-λ^5-phosphorin was formed.

In CCl_4, 2.4.6-tri-tert-butyl-phosphorin 24 absorbed 2 moles of chlorine. Mach [44] isolated a crystalline sterically uniform substance, m.p. 77–78,5 °C, identified as 4-chloro-phosphinic acid chloride 94. 1.1.1.4-Tetra-chloro-2.4.6-tri-tert-butyl-λ^5-phosphorin 93 is assumed to be the primary chlorination product.

94 with sodium methanolate gives the two isomers 84 on alcoholysis and hydrolysis. With sodium ethanolate under carefully controlled conditions the 4-chloroethyl ester 95, m.p. 155 °C, could be isolated as a sterically uniform com-

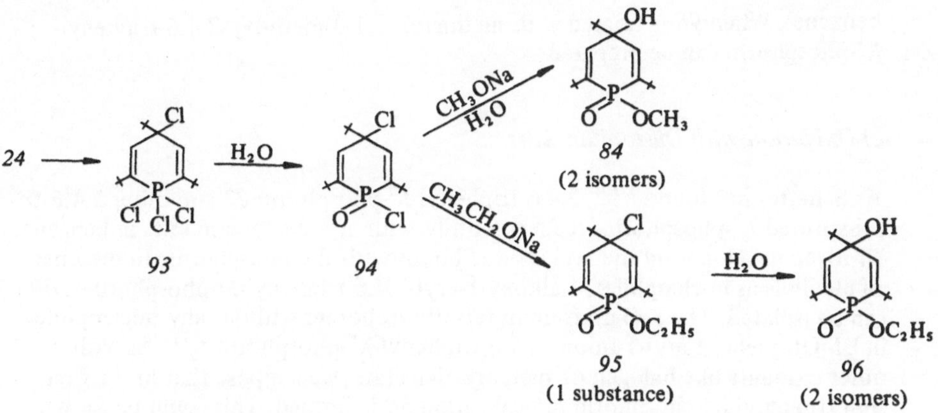

pound. Hydrolysis by $NaOH/CH_3OH$ leads to a *mixture* of the two stereoisomeric hydroxy ethyl esters *96*. Racemization at the phosphorus atom seems less probable than S_N1 type reaction under these conditions.

Recent experiments of Kanter [92] with 2.4.6-triphenyl-λ^3-phosphorine *22* seem very promising. With bromine and light or with pyridine perbromide in the dark, *22* is oxidized to a sensitive solid product *97* which by methanolysis smoothly yields 1.1-dimethoxy-2.4.6-triphenyl-λ^5-phosphorin.

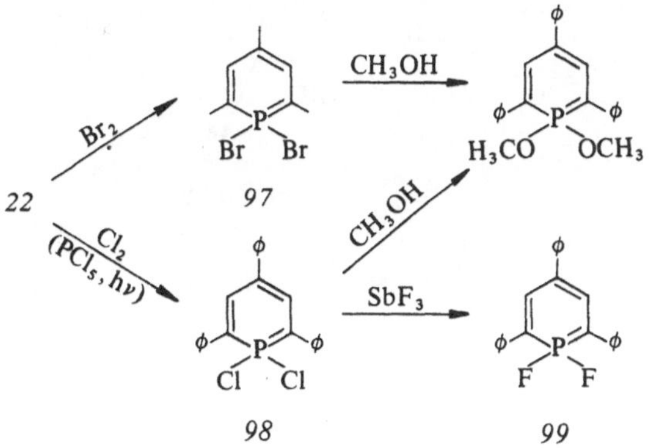

When *22* is treated with phosphorus pentachloride in CCl_4 and light, 1.1-dichloro-2.4.6-triphenyl-λ^5-phosphorin *98*, m.p. 100–3 °C is isolated. ^{31}P NMR in benzene: $\delta = -17,0$ ppm (H_3PO_4, external standard), $J_{P-H} = 50$ Hz (with protons at C–3 and C–5). *98* with SbF_3 yields 1.1-difluoro-2.4.6-triphenyl-λ^5-phosphorin *99*, a very stable yellow substance, m.p. 129–131 °C, ^{31}P NMR: $\delta = -73,0$ ppm ($J_{P-F} = 1041$ Hz; $J_{P-H} = 46$ Hz) (against H_3PO_4 as external standard, benzene) ^{19}F–NMR: $\delta = +47,02$ ppm (CCl_3F as internal standard,

63

benzene). When *98* is treated with methanol, 1.1-dimethoxy-2.4.6-triphenyl-λ^5-phosphorin can be prepared.

e) Oxidation with Diazonium Salts

As Schaffer has found [100], 2.4.6-triphenyl-λ^3-phosphorin *22* and other 2.4.6-tri-substituted λ^3-phosphorins react smoothly with aryl diazonium salts in benzene. Nitrogen develops and the aryl residue bonds with the phosphorus. In presence of alcohols as nucleophiles, 1-alkoxy-1-aryl-2.4.6-triphenyl-λ^5-phosphorins *100* can be isolated. The aryl diazonium-tetrafluoroborate without any nucleophile in DMOE yields 1-aryl-1-fluoro-2.4.6-triphenyl-λ^5-phosphorin *101*. As with other oxidants like halogen or mercury-II-acetate, we suppose that in the first step triphenyl-λ^3-phosphorin radical cation *58* is formed. This could be shown by ESR spectroscopy. The next step may be a radical-radical addition to the λ^4-phosphorin cation or a nucleophile-cation addition to the λ^4-phosphorin radical these than are transformed to the endproducts by a nucleophilic or radical addition respectively:

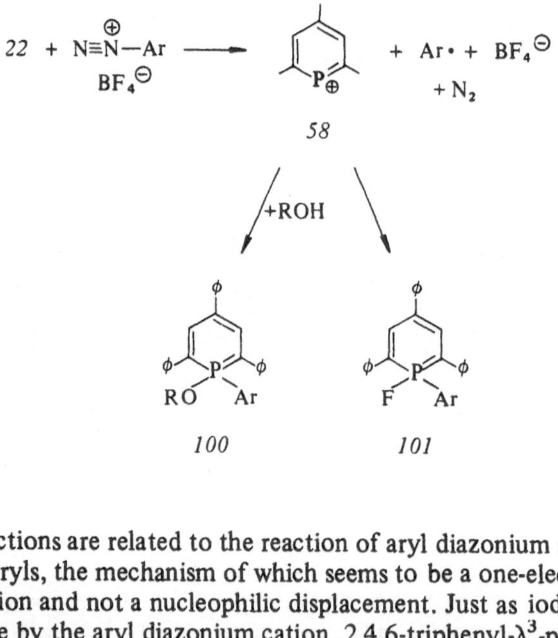

These reactions are related to the reaction of aryl diazonium salts with iodide yielding iodoaryls, the mechanism of which seems to be a one-electron transfer (radical) reaction and not a nucleophilic displacement. Just as iodide is easily oxidized to iodine by the aryl diazonium cation, 2.4.6-triphenyl-λ^3-phosphorin is oxidized to the radical cation *58*.

When 2.4.6-triphenyl-λ^3-phosphorin *22* is treated with an excess of the diazonium salt, for instance, 2-methyl-4-nitro-phenyl-diazonium-tetrafluoroborate *102* in benzene/methanol, a doubly arylated λ^5-phosphorin *103* can be isolated in small amounts (12%). 2.6-Di-tert-butyl-4-phenyl-λ^3-phosphorin 104 leads to 1-aryl-1-methoxy-λ^5-phosphorin. With an excess of the diazonium salt it reacts to 105 by elimination of another mole of nitrogen. It is arylated at position 4' of the aryl ring at C–4 analogously to compound 103.

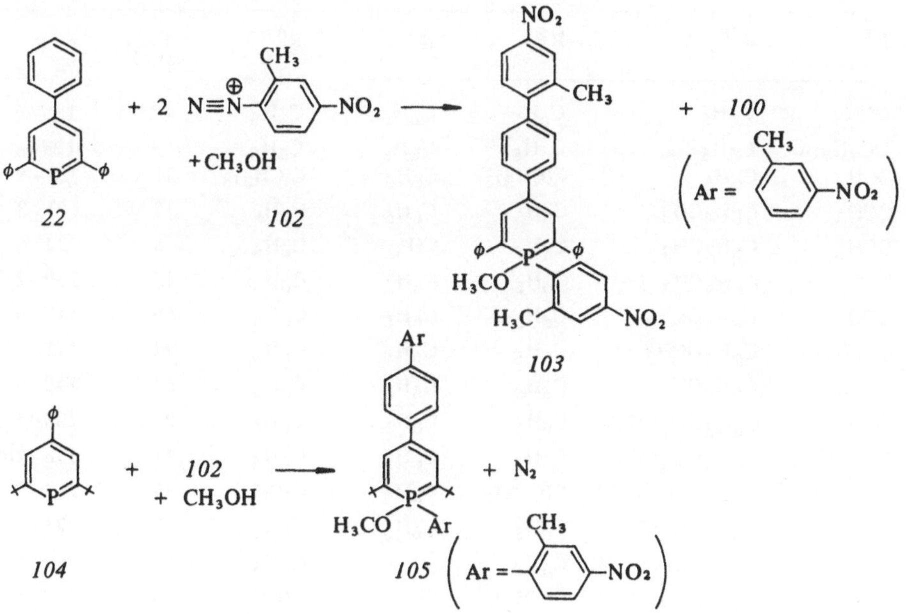

Compounds prepared by these methods are summarized in Table 14. The substance 100 was also prepared by Märkl [88] via pyrylium salt and phenylphosphine, s. Table 23, p. 98.

f) Summary of Oxidation Reactions

As it is clear from the preceding paragraphs, there are four different positions where λ^3-phosphorins are attached by oxidants:

1) Phosphorus: H_2O_2, halogens, diazonium salts lead to 1.1-substituted λ^5-phosphorins or to rearranged 2-hydrophosphinic acids.
2) Phosphorus *and* C–4: oxygen, nitric acid, chlorine in excess. Sensitized oxygen leads to 1.4 addition product 80, oxygen in benzene solution to 4.4'-peroxy phosphinic acid 68. With nitric acid, halogen in excess or autoxidation of 2-

65

Table 14. 1.1-Disubstituted λ^5-phosphorins from λ^3-phosphorins by diazonium salt oxidation

R^1	$R^{1'}$	R^2	R^4	R^6	Yield % [b]	m. p. °C
OCH_3	C_6H_5	C_6H_5	C_6H_5	C_6H_5	47	135–6
$OC_2H_7(n)$	C_6H_5	C_6H_5	C_6H_5	C_6H_5	53	123–4
OCH_3	C_6H_5	$C(CH_3)_3$	C_6H_5	$C(CH_3)_3$	21	137–9
OCH_3	$C_6H_4CH_3$	C_6H_5	C_6H_5	C_6H_5	33	121–3
OCH_3	$C_6H_4CH_3$	C_6H_5	CH_3	C_6H_5	8	134–6
F	$C_6H_4CH_3$	C_6H_5	C_6H_5	C_6H_5	43	130–2
OCH_3	$C_6H_4OCH_3$	C_6H_5	C_6H_5	C_6H_5	59	147–9
OC_6H_{11}	$C_6H_4OCH_3$	C_6H_5	C_6H_5	C_6H_5	24	171–2
OC_2H_5	C_6H_4Cl	C_6H_5	C_6H_5	C_6H_5	64	149–51
OCH_3	$C_6H_3NO_2CH_3$ [a]	C_6H_5	C_6H_5	C_6H_5	23	203–5
OCH_2CH_3	$C_6H_3NO_2CH_3$ [a]	C_6H_5	C_6H_5	C_6H_5	45	198–200
OCH_2CH_3	$C_6H_3NO_2CH_3$ [a]	$C(CH_3)_3$	C_6H_5	$C(CH_3)_3$	35	217–9
F	$C_6H_3NO_2CH_3$ [a]	C_6H_5	C_6H_5	C_6H_5	41	221–3
OCH_3	$C_6H_3NO_2CH_3$ [a]	C_6H_5	[c]	C_6H_5	12	249–51
OC_2H_5	$C_6H_3NO_2CH_3$ [a]	$C(CH_3)_3$	[c]	$C(CH_3)_3$	3,5	237–39

[a] CH_3 in 2′, NO_2 in 4′.
[b] Analytically pure substance, not optimized.
[c]

hydro-phosphinic acids, the 4.4-disubstituted phosphinic acids or their derivatives arise.

3) Phosphorus *and* C–4′ in aryl ring at C–4: aryl diazonium salt in excess.

The cyclic 4-hydroxy-phosphinic acid esters yield diastereoisomers which show remarkable differences in ^{1}H-NMR, in some cases also in IR or in mass spectra.

4. Addition of Carbanions to the P Atom

According to Märkl, Lieb and Merz [77], 2.4.6-trisubstituted λ^3-phosphorins react with lithium or magnesium organocompounds, the carbanionic groups adding at the P center. With 2.4.6-triphenyl-λ^3-phosphorin *22* the deep red delocalized phosphorin anion *106* is initially formed. It is a useful intermediate, reacting with electrophiles at either the P atom or the C–2/C–4 atoms. Addition to the former leads to 1.1-disubstituted λ^5-phosphorins *107*, addition to the latter affords 1.2-dihydro-λ^3-phosphorins *108*, and in some cases 1,4-dihydro-λ^3-phosphorins:

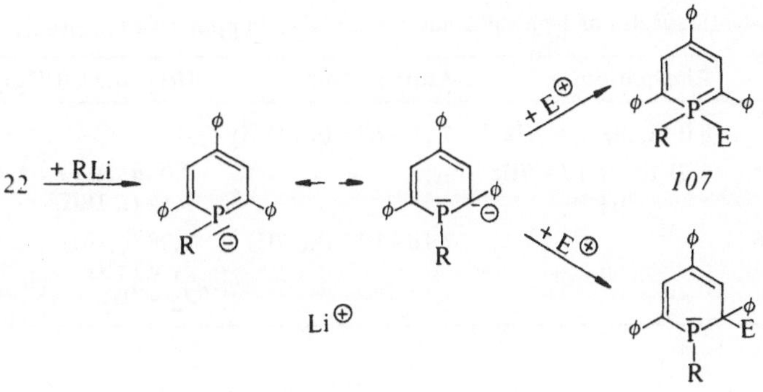

Since there is a close relationship between the classes of compounds *107* and *108*, these reactions will be discussed in detail on page 78.

5. 1.4-Addition

a) Addition of Hexafluoro-butyne-(2)

The diene character of 2.4.6-triphenyl-λ^3-phosphorin *109a* is not well pronounced. It fails to react with acetylene-dicarboxylic acid ester or maleic anhydride and gives with tetracyanoethylene only a charge-transfer band at 450 nm (however, see p. 43). However, Märkl and Lieb [78)] found that *109a* adds the strong electron-accepting hexafluorobutyne-(2) at 100 °C, leading to 1-phospha-barrelene *110a*.

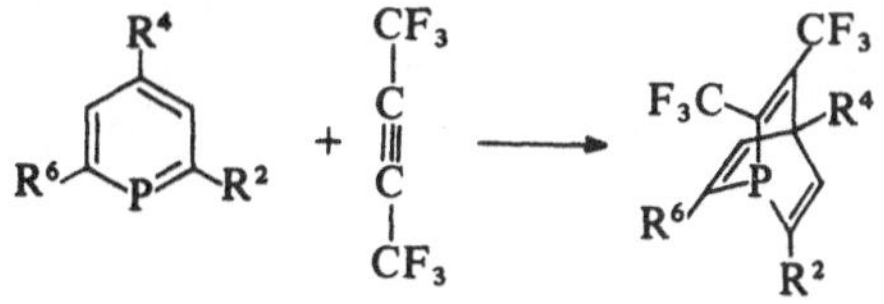

a: $R^2 = R^4 = R^6 = C_6H_5$
b: $R^2 = R^6 = C(CH_3)_3$; $R^4 = CH_3$
c: $R^2 = R^6 = CH_3$; $R^4 = C_6H_5$

67

This addition proceeds more readily with the more electron rich 2.6-di-tert-butyl-4-methyl-λ^3-phosphorin *109 b* or 2,6-dimethyl-4-phenyl-λ^3-phosphorin *109 c*. The ^{1}H-NMR signals are in accord with the proposed structure (Table 15). The ^{31}P resonance of the phospha-barrelene *110 a* in benzene appears at δ = + 65 ppm (H_3PO_4 standard) which corresponds to the ^{31}P resonance of tertiary phosphines.

Table 15. ^{1}H-NMR spectra of 1-phosphabarrelenes *110* (δ in ppm; CDCl$_3$ solvent)

	m. p.	Ring-protons	Arom.protons	CH$_3$- and C(CH$_3$)$_3$-protons
110 a	189°	8.0 (d, 2H) J = 7Hz	7.98–7.12 (m, 15H)	–
110 b	92–3°	6.61 (d, 2H) J = 7Hz	–	2.0 (q, 2H) J = 2.7Hz 1.14 (s, 18H)
110 c	105–6°	–	7.78–7.17 (m, 7H)	2.08 (q, 6H) J_1 = 14Hz J_2 = 2Hz

In a similar way dicyanoacetylene could be added to λ^3-phosphorins [79].

b) Addition of Arynes

Märkl, Lieb and Martin [80] were also able to add arynes *112* to 2.4.6-triphenyl-λ^3-phosphorins; the yields are better with 2.4.6-tri-tert-butyl-λ^3-phosphorin. Here again 1,4 addition takes place with the formation of the 1-phosphabarrelenes *113*. The arynes were generated either from 2-fluorophenylmagnesium bromide or pentachlorophenyl-lithium. The reaction of the more nucleophilic 2.4.6-tri-tert-butyl-λ^3-phosphorin with benzene-diazonium carboxylate also leads to 1,4 addition. The structure of the benzo-phosphabarrelenes *113a–d* is supported by analytical and spectroscopic data (Table 16).

a: R^2 = R^4 = R^6 = C_6H_5;   R = H
b: R^2 = R^4 = R^6 = C_6H_5;   R = Cl
c: R^2 = R^4 = R^6 = C(CH$_3$)$_3$;   R = H
d: R^2 = R^4 = R^6 = C(CH$_3$)$_3$;   R = Cl

Table 16. UV and ^{1}H-NMR spectra of benzo-1-phosphabarrelenes *113*

Compound	m. p.	λ_{max} (nm)	^{1}H-NMR in δ (ppm)	
113 a	207–8 °C	266 (11500) [a]	7.98 (d, 2H) J = 6Hz [d]	
		302 (6900)	7.87–6.28 (m, 19H)	
113 b	246 °C	253 (32500) [b]	8.03 (d, 2H) J = 6.5Hz [e]	
		302 (8300)	7.90–6.94 (m, 15H)	
113 c	108–9 °C	222 (7600) [c]	6.91 (d, 2H) J = 7Hz	1.62 (s, 6H);
				1.28 (s, 3H)
		244 (5100)	7.82–7.78 (m, 4H)	1.13 (s, 18H)
113 d	214 °C	216 (24200) [b]	6.89 (d, 2H) J = 7Hz	1.90 (s, 6H);
				1.23 (s, 3H)
		231 (28300)		1.14 (s, 18H)

[a] In chloroform.
[b] In hexane.
[c] In ethanol.
[d] In deuterochloroform.
[e] In tetrachloromethane.

Attempts to thermally split off acetylenes from *113* in the hope of forming phosphanaphthalenes failed.

6. Reactions with Carbenes or Carbenoids

In view of the smooth addition of carbanions to the electrophilic P atom of λ^3-phosphorins (see p. 78 and 66), Märkl and Merz [81] attempted the addition of carbenoids (carbenes?) by reacting dichloromethane, trichloromethane or dichlorophenylmethane with potassium-tert-butoxide in the presence of 2.4.6-triphenyl-λ^3-phosphorin *22* or 2.4.6-tri-tert-butyl-λ^3-phosphorin *24*. The desired bicyclocompound *115* or the λ^3-phosphepin *116* were not obtained. Instead, benzene derivatives were formed by loss of a "PCl" fragment, the fate of which was not determined. Märkl has proposed a mechanism in which intermediates *114, 115* and *116* are invoked:

or *22* (R = φ)
24 (R = +)

114   *115*   *116*   *117*

R′ = H, Cl or C_6H_5

IV. λ^5-Phosphorins

A. Introduction and Review

As early as 1963 Märkl [82] prepared the first representatives of this class of compounds via multi-step synthesis: 1,1-diphenyl-λ^5-phosphorin *118* ("1,1-diphenyl-phosphabenzene") and 1,1-diphenyl-2,3-benzo-λ^5-phosphorin *120* ("1,1-diphenyl-phospha-naphthalene"). Neither compound could be obtained in crystalline form. Instead, treatment of the crystalline phosphonium salts *119* and *121* with aqueous alkali affords very reactive, air-sensitive yellow or orange powders (*118* and *120*). Acid treatment leads back to the phosphonium salts.

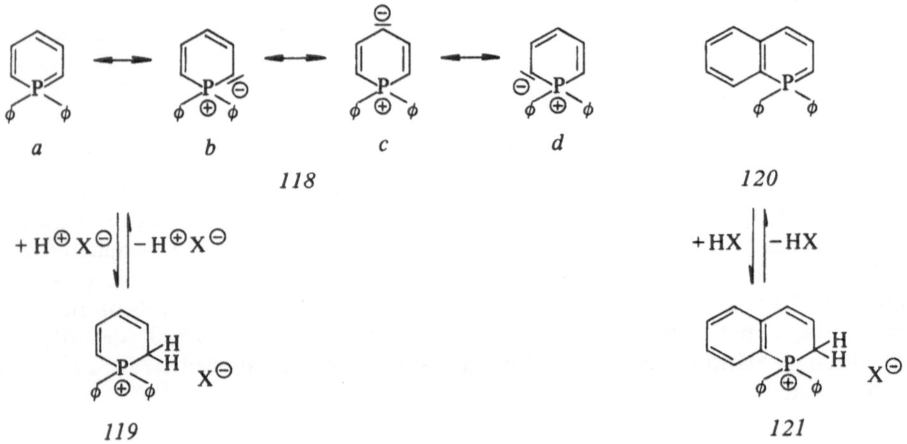

The physical and chemical properties of the λ^5-phosphorins *118* and *120* are comparable to those of phosphonium ylids which are resonance-stabilized by such electron-pulling groups as carbonyl or nitrile substituents [83]. Thus they can be viewed as cyclic resonance-stabilized phosphonium ylids (*118* b, c, d). As expected, they do not react with carbonyl compounds giving the Wittig olefin products. However, they do react with dilute aqueous acids to form the protonated salts. Similarly, they are attacked at the C−2 or C−4 positions by alkyl-, acyl- or diazonium-ions [84]. Heating with water results in hydrolytic P−C cleavage, phosphine oxide and the hydrocarbon being formed.

70

Nevertheless, the stabilization of the ylid by the cyclic delocalized 6π-electron system presents some interesting theoretical problems. In this connection Märkl [85, 86] has coined the term *"non-classical phosphabenzene"*. We will return to this point (p. 115).

In contrast to the unsubstituted ring compounds *118* and *120*, 1,1-diaryl-, 1,1-dialkyl- or 1-aryl-1-alkyl-λ^5-phosphorins with three phenyl groups in positions 2,4 and 6 of the λ^5-phosphorin ring *122* are much easier to prepare and to handle. They can be obtained either from 2.4.6-triphenyl-pyrylium salts or from 2.4.6-triphenyl-λ^3-phosphorins. Most of these 2.4.6-tri-substituted λ^5-phosphorins are very stable and can be isolated as well-defined crystalline compounds. They do not react with the above-mentioned cations. However, reversible protonation-deprotonation does take place in the presence of acids.

122   *123*

R^1, $R^{1'}$ = aryl, alkyl

Table 17. Types of 1.1-hetero- and 1-hetero-1-carbo-λ^5-phosphorins *125*

x	y	x	y
O-Alkyl	O-Alkyl	$N(Alkyl)_2$	$N(Alkyl)_2$
O-Alkyl	O-Aryl	$N(Aryl)_2$	$N(Aryl)_2$
O-Aryl	O-Aryl	$(Alkyl)N$-$(CH_2)_n$-$N(Alkyl)$ ($n = 2,3$)	
O-$(CH_2)_n$-O ($n = 2,3,4$)		S-Alkyl	S-Alkyl
		S-$(CH_2)_n$-S ($n = 2,3$)	
		S-Alkyl	$N(Alkyl)_2$
O-Alkyl	O-Acyl	$N(Alkyl)_2$	F
O-Acyl	O-Acyl	F	F
O-Alkyl	O-Z[*)]	Cl	Cl
O-Alkyl	O-Z'[**)]	Br	Br
O-Alkyl	$N(Alkyl)_2$	O-Alkyl	Alkyl
O-Alkyl	$N(Aryl)_2$	O-Alkyl	Aryl
O-$(CH_2)_2$-NH		O-Aryl	Alkyl
O-$(CH_2)_2$-N(Alkyl)		O-Aryl	Aryl
		F	Aryl

[*)] Z =   [**)] Z' =

By preparing 1.1-dialkoxy-λ^5-phosphorins *124*, Dimroth and Städe [61] obtained the first representatives of new classes of 1.1-heterosubstituted λ^5-phosphorins 124. In contrast to the 1.1-carbosubstituted "1.1-carbo-λ^5-phosphorins" *122*, we call these "1.1-hetero-λ^5-phosphorins" *125*. Table 17 summarizes the known types together with some mixed 1-carbo-1-hetero-λ^5-phosphorins.

R^2, R^4, R^6 = alkyl or aryl
R, R' = alkyl or aryl
X, Y = O, N, S, C-residues or Br, Cl, F

124   *125*

There are a number of different syntheses. The 1.1-hetero-λ^5-phosphorins lead to many new and unexpected reactions.

B. Synthesis

1. General Remarks

The unsubstituted compounds *118* and *120* must be prepared via a large number of steps involving simple building blocks. This method has little preparative value. Moreover, the unsubstituted λ^5-phosphorins appear to be so reactive that they are rather difficult to handle. Thus, the emphasis in synthesis has been placed on the preparation of 2.4.6-tri-substituted derivatives, which will be dealt with exclusively in this section.

Three primary synthetic sequences have been developed:
  1) Treatment of pyrylium salts with aryl-(or alkyl)-phosphines or their bis-hydroxymethyl derivatives.
  2) Transformation of λ^3-phosphorins to λ^5-phosphorins.
  3) Transformation of λ^5-phosphorins to other λ^5-phosphorin-derivatives.

Methods 2) and 3) allow for the highest degree of synthetic variation. λ^3-Phosphorins differ considerably in their chemical properties from the pyridine analogs. This contrasting behavior can be traced to the following three points:
  1) The smaller electronegativity of P compared to N.
  2) The possibility of P expanding its valence shell beyond 8.
  3) The different orbital energies of the P and N ring systems (see p. 37).

126   *127*   *128*

The fact that λ^3-phosphorins are not basic, as discussed above, matches the finding that *phosphorinium ions 127*, corresponding to the well-known pyridinium ions *126*, cannot be isolated. However, they play a key role as *intermediates* in many syntheses, since they are easily attacked by nucleophiles, forming λ^5-phosphorins *128*.

The most important synthetic sequences leading to the actions *127* are summarized in the following five points:

1) Treatment of 2.4.6-tri-substituted pyrylium salts with phosphines (C. C. Price, Märkl).
2) Oxidation of λ^3-phosphorins with hydrogen peroxide (Dimroth and Vogel).
3) Oxidation of 1-substituted phosphorin radicals with mercuric acetate (Dimroth and Hettche).
4) Oxidation of 1-substituted 1,2-dihydro-λ^3-phosphorins with mercuric acetate or triphenylcarbonium salts (Märkl).
5) Treatment of λ^3-phosphorins with diazonium salts (Dimroth and Schaffer). This reaction may also have a radical cation *58* as intermediate (s. p. 64).

On the other hand, Märkl has found that λ^3-phosphorins add lithium or magnesium compounds at the P atom, forming stable 1-aryl-or 1-alkylphosphorin anions

128

133. In pyridine chemistry the analogous addition does not lead to an attack at the nitrogen atom *130*, but rather to addition at the C—2 or C—4 atoms of the ring, *131* and *132*.

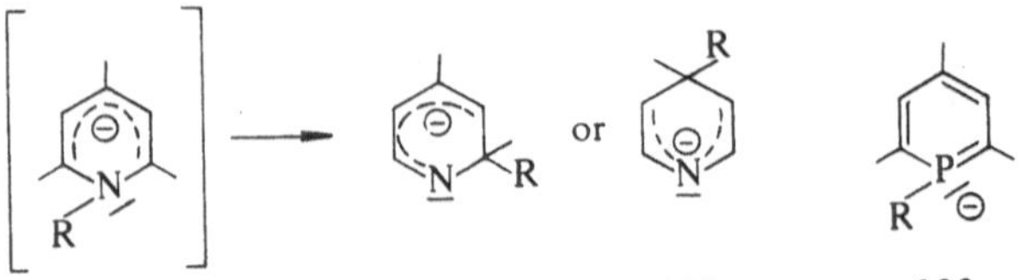

Besides this method (1) of adding nucleophiles, anions *133* can also be prepared by: (2) reduction of λ^3-phosphorins to radical anions *134* followed by reaction with radicals, and (3) addition of radicals to the P atom, forming phosphorin radicals *135* followed by reduction.

Methods (2) and (3) have not been explored from a preparative viewpoint.

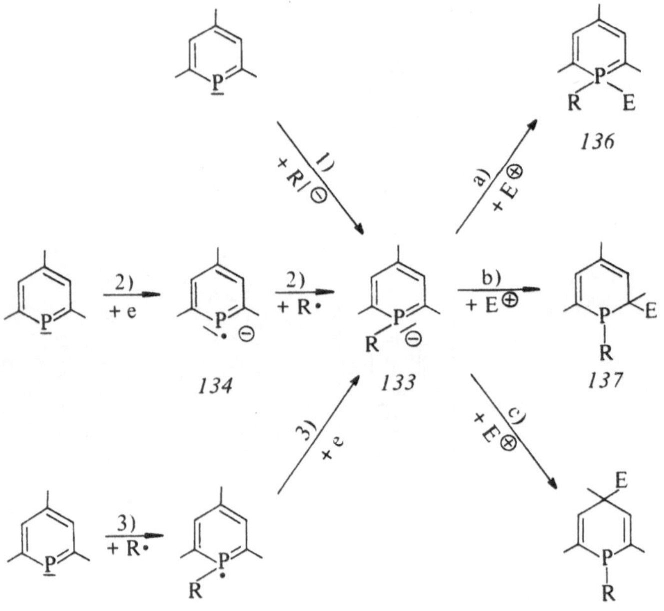

The P-substituted phosphorin anions *133* are ambident. The site of electrophilic addition depends upon the reaction conditions:

a) At the P atom to form λ^5-phosphorins *136*
b) At C−2 to form 1,2-dihydro-λ^3-phosphorins *137*
c) At C−4 to form 1,4-dihydro-λ^3-phosphorins *138*

A third reaction path, which again has no analog in pyridine chemistry, involves the stable λ^3-phosphorin cation radical *58* as an intermediate. It is easily formed by oxidation of λ^3-phosphorins:

If the oxidation is carried out with a radical R in the presence of a nucleophile $R^\ominus$, an equilibrium reaction forming the neutral radical *135* may take place. *135* could lead to the λ^5-phosphorin *129* via oxidation to *127* and addition of the anion $R'^\ominus$, or by simple coupling with the radical R'.

Not all of the above-discussed reaction paths have been carried out or preparatively explored and very few have been studied in detail mechanisticly.

The versatile reactions of λ^3-phosphorins and λ^5-phosphorins are highly influenced by the nature of the substituents at the ring and above all by the bond energy and eletronegativity of the P substituents. Thus, under certain conditions an exchange of P substituents in λ^5-phosphorins can be induced. Similarly, a rearrangement from the P atom to position C−2 in the ring, or even a complete cleavage of the substituents is conceivable.

Finally, we point to the possibility of P = O bond formation from 1-alkoxy-λ^5-phosphorin derivatives *124* or *125* by cleavage of alkyl cations. Also the reverse process, *i. e.* alkylation of the P = O moiety to form P−O−R groups is possible. The synthesis of λ^5-phosphorins having functional groups at the C-atoms of the phosphorin ring was first made possible by the preparation of new stable λ^5-phosphorin carbenium ions *140*. Here again, the fundamental difference between phosphorin and pyridine systems comes to light: Whereas *carbanionic* structures *139 b* are stabilized in the pyridine series, in the λ^5-phosphorin series *carbenium* ions as *140 b* are stabilized.

In other words: Whereas the chemistry of pyridine is similar to that of nitroben-zene, both being electron-deficient benzene derivatives, the chemistry of the λ^3-phosphorins, and particularly that of the 1.1-hetero-λ^5-phosphorines, is similar to that of N.N-dimethylaminobenzene, both being electron-rich benzene derivatives.

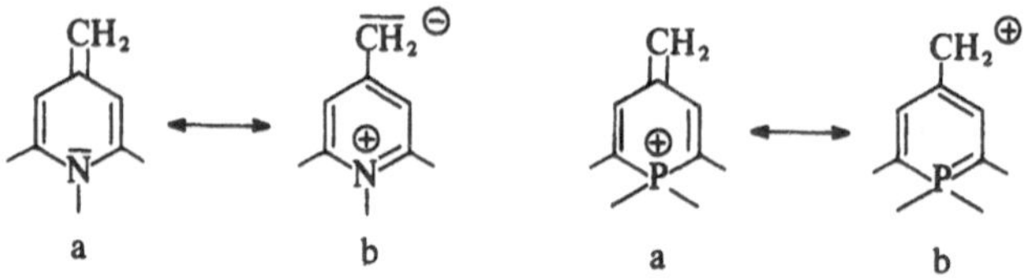

In summary, the peculiarity of phosphorin chemistry is based on the unique participation of phosphorus in a "classical" or "non-classical" 6 π-electron system. The different types of valence and bonding which are characteristic of phosphorus, as well as d-orbital participation, are the most interesting and important factors.

2. Specific Methods of Synthesis

a) 1.1-Carbo-λ^5-phosphorins: Dialkyl-, Diaryl- or Alkylaryl-λ^5-phosphorins

b) Method A: Utilization of Saturated Cyclic Phosphorus Compounds

In the preparation of 1,1-diphenyl-λ^5-phosphorin, Märkl [82] used the following sequence:

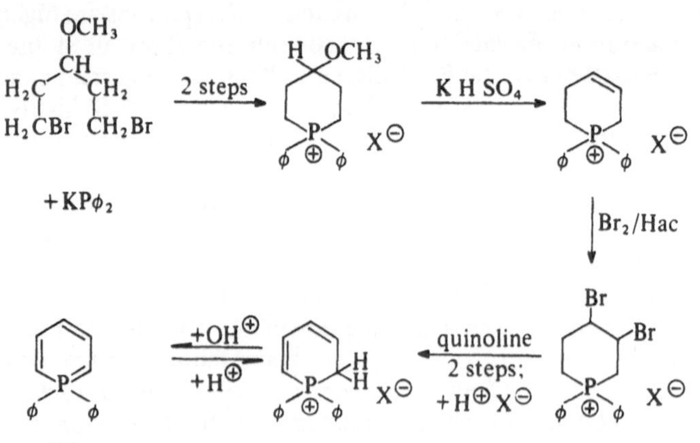

118 could be obtained only as a yellow powder ($\lambda_{max} = 409$ nm, in CH_3OH), oxidizing in the presence of air to compounds which range in color from red to violet; the oxidation products were not identified. Acids cause reconversion to the phosphonium salt 119. The perchlorate (119 (X = ClO_4)) crystallizes from aceto-nitrile with one mole of solvent (m. p. 175–6 °C) and has IR-absorption bands at 1626, 1572 and 1550 cm^{-1}. The parent compounds of this series, 1,1-dihydro-λ^5-phosphorin or 1,1-dimethyl-λ^5-phosphorin (118, H or CH_3 in place of C_6H_5, respectively) are not known.

The 1,1-diphenyl-λ^5-phosphorins 120 a–c are colored, easily autoxidizable, non-crystalline compounds which were prepared by multistep synthesis, the last step being conversion of the corresponding 1.1-diphenyl-2.3-benzophosphonium per-chlorates 121 a–c by aqueous alkali.

a: R = H	$\lambda_{max} = 420$ nm (benzene)	a: R = H	m.p. 173 °C
b: R = CH_3	$\lambda_{max} = -$	b: R = CH_3	m.p. 197–9 °C
c: R = C_6H_5	$\lambda_{max} = 488$ nm (benzene)	c: R = C_6H_5	m.p. 182–3 °C

These λ^5-phosphorins 120 a–c also fail to react with carbonyl compounds. However, they are attacked by electrophiles ($H^\oplus$ or alkyl cations) at the C–2 position. In this manner new 1,1-diphenyl-2,3-benzo-λ^5-phosphorins which are sustituted at positions C–2 (and C–4) can be prepared. Diazonium ions attack at C–4 to form azocompounds; if an excess is used, C–2 is also substituted [32, 82b)]. Hydrolysis with hot water affords 141. The reaction with ortho-formic acid ester forms a cyanine dye having a bridge at the C–4 positions 142 [32)]. The experimen-tal details have not yet been published.

141

142

c) Method B: Reaction of 2.4.6-Trisubstituted λ^3-Phosphorins with Organometallic Compounds Followed by Treatment with Alkyl- and Acylhalides

Märkl, Lieb and Merz [77)] have described the carbanionic addition of lithium or magnesium organometallic compounds to the P atom of 2.4.6-triphenyl-λ^5-phosphorins, which form deep red salts *143*. These can be alkylated either at the P atom to form λ^5-phosphorins *144* or at C–2 to yield 1,2-dihydro-λ^3-phosphorins *145*. Acylation with benzoylchloride affords the 1,4-dihydro-λ^3-phosphorin derivatives *146*. Addition of acids or water leads to the synthetically important intermediates *147* which can be reconverted to the λ^5-phosphorin-salts *143* by 2 N NaOH.

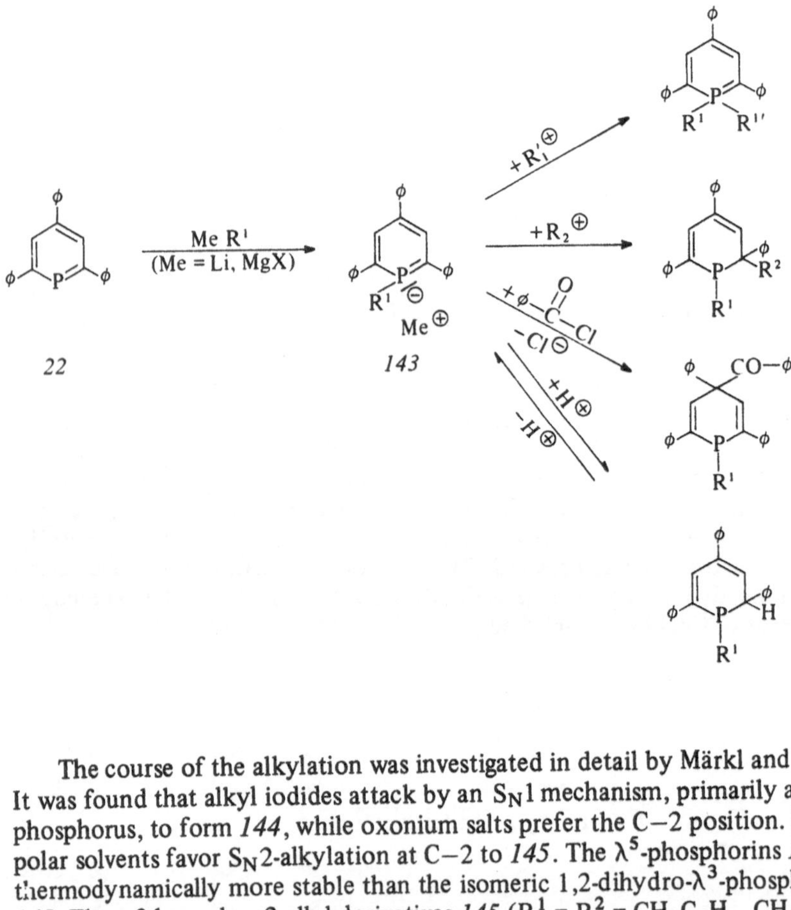

The course of the alkylation was investigated in detail by Märkl and Merz [87)]. It was found that alkyl iodides attack by an S_N1 mechanism, primarily at the phosphorus, to form *144*, while oxonium salts prefer the C–2 position. Nonpolar solvents favor S_N2-alkylation at C–2 to *145*. The λ^5-phosphorins *144* are thermodynamically more stable than the isomeric 1,2-dihydro-λ^3-phosphorins *145*. Thus, 2-benzyl or 2-allyl derivatives *145* ($R^1 = R^2 = CH_2C_6H_5$, CH_2–CH = CH_2) rearrange at 180–220° to the 1,1-compounds *144*. At higher temperatures these substituents are split off, forming 2.4.6-triphenyl-λ^3-phosphorin *22*. Attempts to convert 2.4.6-tri-tert-butyl-λ^3-phosphorin to 1,1-dimethyl-2.4.6-tri-tert-butyl-λ^5-phosphorin by this route led only to the 1,2 derivative ($R^2 = CH_3$) corresponding to *145* (+ instead ϕ) (m. p. 98 °C) [88].

d) Method C: Oxidation of 1.2-Dihydro-λ^3-phosphorins with Triphenyl-methyl-perchlorate Followed by Reaction with Phenyl-lithium

Oxidation of dihydro-λ^3-phosphorin *147* with triphenyl-methyl-perchlorate leads to an intermediate which most likely has the structure *127*. Addition of phenyl-lithium to this cation affords the deep red 1,1,2,4,6-pentaphenyl-λ^5-phosphorin *144* [86].

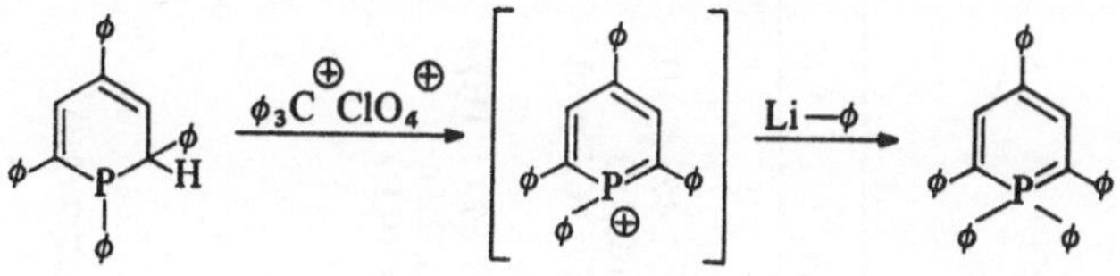

e) Method D: Treatment of λ^3-Phosphorins with Organomercury Compounds

In analogy to the reaction discussed on page 84 (Method G), Märkl was able to convert 2.4.6-triphenyl-λ^3-phosphorin *22* to 1,1-diaryl-2.4.6-triphenyl-λ^5-phosphorin *144* with diarylmercury compounds at 240–260 °C [86]. The radical *135* can be assumed to be an intermediate, since the 1,2-dihydro compound *147* also reacts with diaryl-mercury derivatives to form the same end product *144* at temperatures as low as 220 °C.

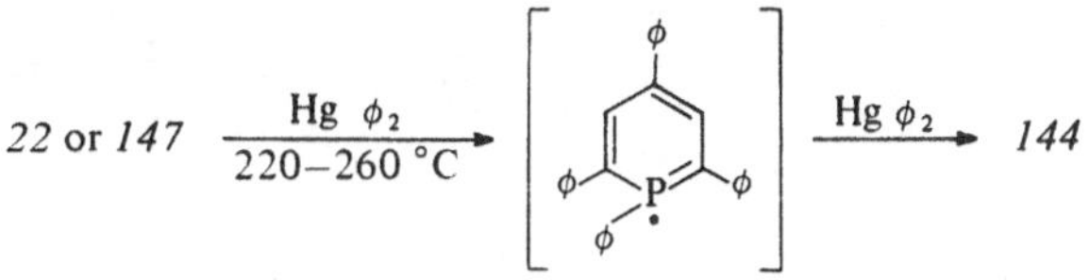

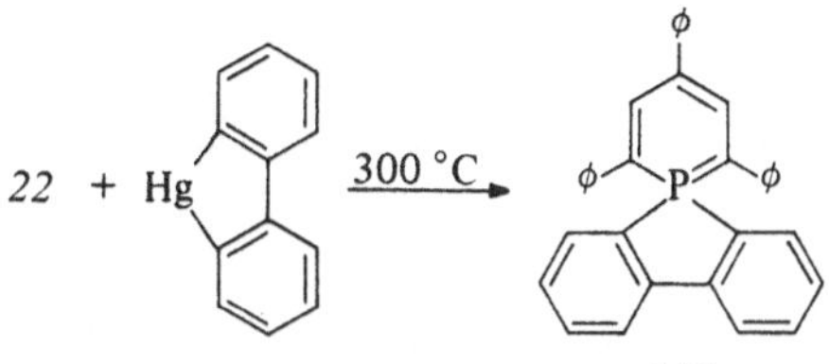

The same procedure can be used to convert *22* to the spiro compound *148* using diphenylenemercury at 300 °C.

Tables 18 and 19 summarise known compounds of type *144, 145* and *146*.

Table 18. 1.1-Dialkyl-, diaryl and alkyl-aryl-2.4.6-triphenyl-λ^5-phosphorins *144*

R^1	$R^{1'}$	m. p. [°C]	λ_{max} nm (ϵ)	^{1}H NMR (δ in ppm)[b]	Lit.
$CH_2C_6H_5$	$CH_2C_6H_5$	Oil	515		87)
C_6H_5	CH_3	169–70°	518 (9300)[a] 398 (2400) 335 (18000)	7.6–6.75 (m, 6H); 7.42 (d, 1H) $J = 37.5$ Hz; 2.07 (d, 3H) $J = 13$ Hz.[b]; δ^{31}P = +6.5 ppm (gegen H_3PO_4)	85)
C_6H_5	C_2H_5	151–2°	513 (8900)[a] 400 (2390) 335 (21800)	7.8–6.7 (m, 6H); 7.62 (d, 1H) $J = 41$ Hz; 2.51 (m, 2H); 1.03 (2 tripl.) $J_1 = 19$ Hz, $J_2 = 7$ Hz.[b]	85)
C_6H_5	CH_2-C_6H_5	207–8° (201–3)	514 (8500)[a] 394 (2350) 345 (16300)		87, 77)
C_6H_5	CH_2-CH = CH_2	Oil	515		87)
C_6H_5	C_6H_5	191–3°	515 (8900) 342 (17000)	7.8–6.9 (m, 25H); 7.7 (2H) $J = 34$ Hz.[b]	86)
$C_6H_4CH_3$	$C_6H_4CH_3$	225–8°	516 (10500) 340 (18600)	7.7–6.85 (m, 23H); 7.63 (2H) $J = 34$ Hz; 233 (6H)	86)

Table 18 (continued)

R¹	R¹'	m.p. [°C]	λ_{max} nm (ϵ)	¹H NMR (δ in ppm)[b]	Lit.
C₆H₄Br	C₆H₄Br	251–4°	519 (9900), 337 (17400)	7.75–7.0 (m)	86)
α-Naphthyl	α-Naphthyl	260–2	525 (10500), 346 (42700)	7.95–6.9 (m)	86
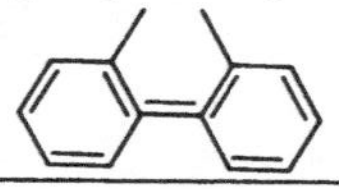 (2,2'-dimethylbiphenyl)		277–80	474 (6300)[c], 336 (11300)		86)

a) In Cyclohexan.
b) In CDCl₃.
c) In Benzol.

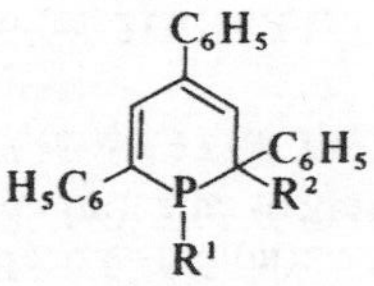

Table 19. 1.2-Dialkyl-, diaryl- and alkyl-aryl-2.4.6-triphenyl-1.2-dihydro-λ^3-phosphorin *146* and 1.4-dihydro-λ^3-phosphorin *146*

C_6H_5, C_6H_5, R^2, P, R^1, H_5C_6 *145*

R¹	R²	m.p.	λ_{max} (nm) (ϵ)	¹H NMR (δ in ppm)[b]	Lit.
CH₃	H	95–96°	320	7.9–7.0 (m, 16H); 6.25 (m, 1H); 6.25 (m, 1H); 0.75 and 1.13 (quart.).	87)
n-C₄H₉	H	Oil		7.9–6.9 (m, 16H); 6.22 (d, 1H); 6.04 (quart. 1H) J_1 = 8 Hz J_2 = 4 Hz; 1.5–0.4 (m, 9H)	77, 87)
CH₂C₆H₅	H	145–6°	330 (8700), 225 (21400)	7.85–6.75 (m); 5.96 (d); J = 2.5 Hz; 4.21 (quart.) J_1 = 12.5 Hz, J_2 = 4Hz.	87)

Table 19 (continued)

R¹	R²	m. p.	λ_{max} (nm) (ϵ)	¹H NMR (δ in ppm)[b]	Lit.
C_6H_5	H	144–5°	322 (6750) 255 (20000)	7.7–6.7 (m, 1H); 4.14 (quart., 1H) $J_1 = 3.5$, $J_2 = 7$ Hz. 3.74 (quart. 1H) $J_1 = 2.5$ Hz, $J_2 = 7$ Hz; 2.92 (d) $J = 5$ Hz.	77, 87)
C_6H_5	CH_3	147–8°	322 (91200)[a] 354 (25100)	8.0–6.96 (m, 21H); 6.25 (d, 1H) $J = 2.5$ Hz; 1.78 (d, 3H) $J = 18$ Hz	87)
C_6H_5	C_2H_5	184–5°	325 (8100)[a] 254 (24500)	7.82–6.67 (m, 21H); 6.28 (d, 1H) $J = 2$ Hz; 2.26 (m, 2H); 0,95 (tripl., 3H) $J = 7$ Hz.	85, 87)
C_6H_5	CH_2-CH = CH_2	155–6°	323 (9700) 253 (38000)	7.55–6.75 (m); 6.21 (d) $J = 2.2$ Hz; 6.0–5.5 (m); 5.26–4.83 (m); 3.03 (octett)	85)
C_6H_5	$CH_2C_6H_5$	168–9°	324 (8900) 253 (22400)	7.78–6.7 (m); 5.97 (d) $J = 2$ Hz; 3.61 (octett).	85) 87)
$C_6H_4N(CH_3)_2$	H	146–7°	333 (17000) 267 (34000)	7.96–6.6 (m, 21H); 6.14; 4.00 (quart. 1H); 2.88; 2.86 (2s, 6H).	87)
CH_2-C_6H_5	$CH_2C_6H_5$	145–6°	330 (8700) 253 (22400)	—	87)
$CH_2C_6H_5$	H *)	129–31	319 (7100) 261 (19500)	7.6–6.9 (m, 21H); 6.4–6.2 (m, 1H); 3.0–2.3 (m, 3H); (1.28, 9H)	87)

*) For C_6H_5 at C–4: $C(CH_3)_3$.
a) In ethanol.
b) In CDCl₃.
c) In cyclohexane.
d) In benzene.
e) Diastereoisomer.

f) 1.1-Hetero-λ^5-phosphorins and 1-Carbo-1-hetero-λ^5-phosphorins

g) Method E: The Action of Radicals on λ^3-Phosphorins

Addition of the double molar amount of 2.4.6-triphenylphenoxyl 57 in benzene to 2.4.6-triphenyl-λ^3-phosphorin 22 leads to the formation of deep yellow crystals of 1.1-bis-(2.4.6-triphenylphenoxy)-2.4.6-triphenylphosphorin 149 in approx 30%

yield [45]. The radical cation *58* can be identified as an intermediate by ESR spectroscopy. *58* seems to be in equilibrium with the neutral radical *59*.

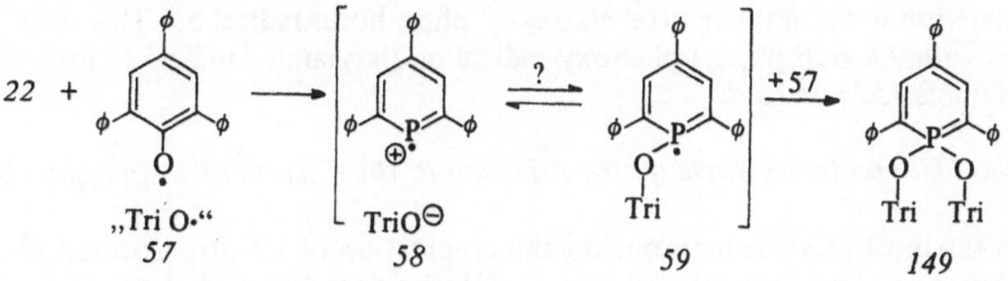

Hettche [88, 94] was able to add diarylaminyls, which can be generated by thermolysis of hexa-aryl-hydrazines *150* [91], to 2.4.6-triphenyl-λ^3-phosphorin *22* at ca. 80 °C, forming 1,1-bis-diarylamino-λ^5-phosphorin *151*. The analytically pure λ^5-phosphorins are obtained in 60–65% yields.

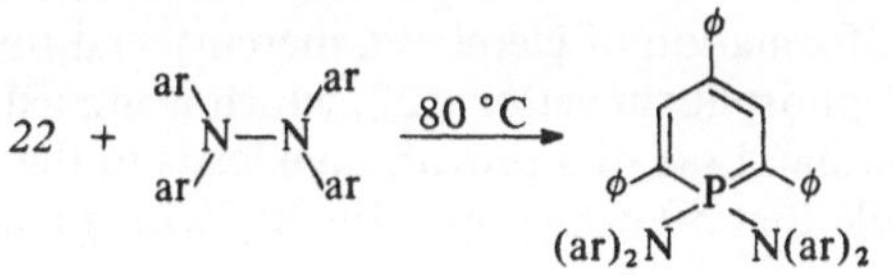

The course of this reaction should be similar to the triphenylphenoxyl addition.

h) Method F: Simultaneous Reaction of Radicals and Nucleophiles with λ^3-Phosphorins

If the radical reactions (Method E) with triphenylphenoxyl or diphenylaminyl are carried out in the presence of an excess of alcohols, one obtains the mixed 1,1-substituted λ^5-phosphorin derivatives with either phenoxy or diarylamino and one

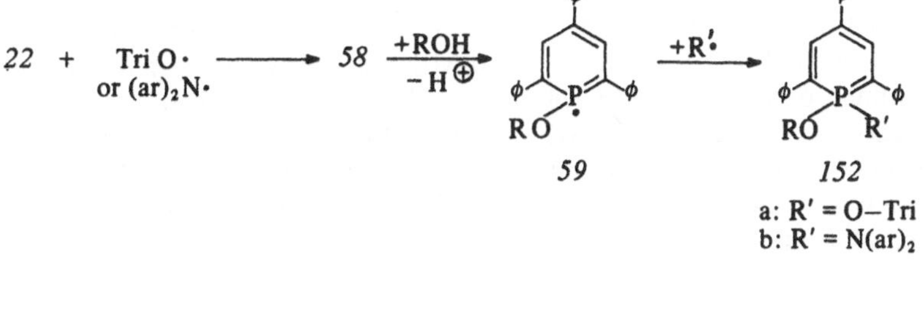

a: R' = O–Tri
b: R' = N(ar)$_2$

alkoxy group. The intermediate in these reactions is probably the 2.4.6-triphenyl-phosphorin radical cation *58*. This preferentially reacts, especially under basic conditions with the more basic and abundant alcoholate ion, instead the less nucleophilic, steric-hindered, weakly nucleophilic 2.4.6-triphenylphenol or phenolate ion to form the neutral alkoxy-λ^4-phosphorin radical *59*. This then couples with 2.4.6-tri-phenyl-phenoxy radical or diarylamine radical to form the products *152a* or *152b*.

i) Method G: Oxidative Nucleophilic Addition to the P Atom of λ^3-Phosphorins

One of the most versatile methods for the preparation of 1,1-disubstituted λ^5-phosphorins *124* was discovered by Städe [45] who found that λ^3-phosphorins *2* can be oxidized (mercuric acetate gives the best results) in the presence of alcohols or phenols in benzene to 1.1-dialkoxy- or 1.1-diphenoxy-λ^5-phosphorins *124*. The first step is probably a reaction of the „soft" λ^3-phosphorin- π-system with the „soft" acid $Hg^{2\oplus}$ which by electron transfer leads to the weakly electrophilic radical cation *58*. This is then attacked by alcohol or phenol – or as Hettche has found [88] by other nucleophiles such as an amine to form by loss of a proton the neutral λ^4-phosphorin radical *59*. This radical is oxidized once again by mercury ions leading to the formation of elemental mercury and the strongly electrophilic, short-lived λ^4-phosphorin cation *127*, which is immediately attacked by alcohol, phenol or amine. Loss of a proton then leads to the λ^5-phosphorin *124*. It is also conceivable that *59* can couple directly with a radical to form *124* (Method E, p. 82).

This reaction can be carried out in good yield with a wide variety of alcohols, phenols and, according to Hettche [88], also with secondary and primary amines. Diols, diamines, or aminoalcohols lead to 1,1-spiro-λ^5-phosphorins.

The reactions are also smooth with 2- or 4-alkyl-substituted λ^3-phosphorins; 1.1-dialkoxy- or 1.1-bis-dialkylamino-4-methyl-2.6-diphenyl-λ^5-phosphorins, which can be readily obtained from 2.6-diphenyl-4-methyl-λ^3-phosphorin, are very useful for further synthesis (see p. 87).

By reacting 2.4.6-triphenyl-λ^3-phosphorin with diphenylmercury at high temperatures, Märkl was able to induce addition of two carbon groups to the phosphorus (Method D, p. 79).

Since the radical cation *58* and the cation *127* are of different electrophilicity, Hettche [88, 98] was able to steer the course of the reaction in such a way that λ^5-phosphorins *153* and of type *154*, which originally formed as byproducts, could be obtained in useful amounts.

153   154   155

$X^1 = X^2 = $ O-alkyl or N(alkyl)$_2$
$X^1 = $ O alkyl: $X^2 = $ N(alkyl)$_2$

The dibenzoate *155* was prepared by Constenla [101]. All of these λ^5-phosphorin proved to be synthetically and theoretically very interesting compounds.

Another very versatile reaction, leading to 1-carbo-1-hetero-λ^5-phosphorins, was found by Schaffer [100] when he reacted λ^3-phosphorins with aryl-diazonium salts in the presence of alcohols. This reaction proceeds probably by an analogous mechanism. It is described on p. 64.

j) Method H: Oxidative Nucleophilic Addition Starting from 1.2-Dihydro-λ^3-phosphorins

If 1-alkyl- or 1-aryl-1,2-dihydro-λ^3-phosphorins *147* (obtained by Method B, p. 78) are oxidized with mercuric acetate in the presence of alcohols or phenols as nucleophiles, it is possible to isolate λ^5-phosphorins *156* in which the phosphorus bears a carbon substituent besides an alkoxy or phenoxy group (Märkl [90]).

Here again the cation *127* is probably an intermediate.

147   156

(R' = alkyl, aryl)

k) Method I: Reaction of Pyrylium Salts with Primary Phosphines

Märkl, Merz and Rausch [90] have repeated the work of Price [22] in which 2.4.6-triphenylpyrylium salts *21* are condensed with phenylphosphine or bis-hydroxymethyl-phenylphosphine in the presence of such nucleophiles as water, alcohols or thiols. 1-Carbo-1-hetero-substituted λ^5-phosphorins *157* are the endproducts. Here again it is reasonable to assume the intermediacy of a cation *127*:

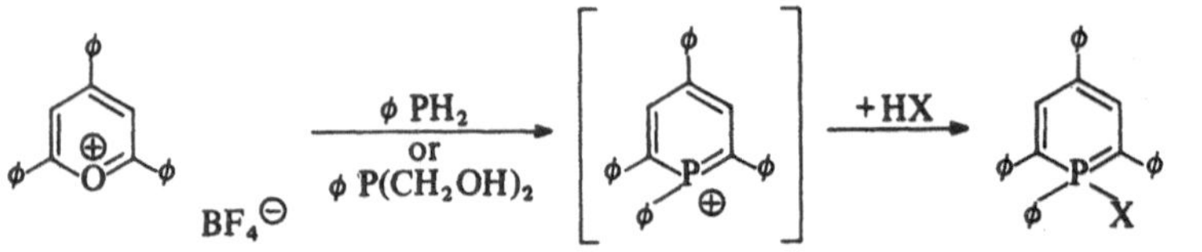

21 *127* *157*
(X = OR, SR)

Price obtained in the presence of water a crystalline product (m. p. 256–257 °C) and an amorphous material from which Märkl [86] could isolate the λ^5-phosphine oxide *158* (m. p. 156–158 °C) and the 4-hydroxy-phosphine oxide *160* (m. p. 239–241 °C) which probably was formed by autoxidation. *158* is also formed by H_2O_2-oxidation of *147* [77]. The tautomeric form of *158* would be the 1-hydroxy-λ^5-phosphorin *159*. Indeed, treatment with base affords a bright red anion which probably has the structure *161* (see p. 60 and 87).

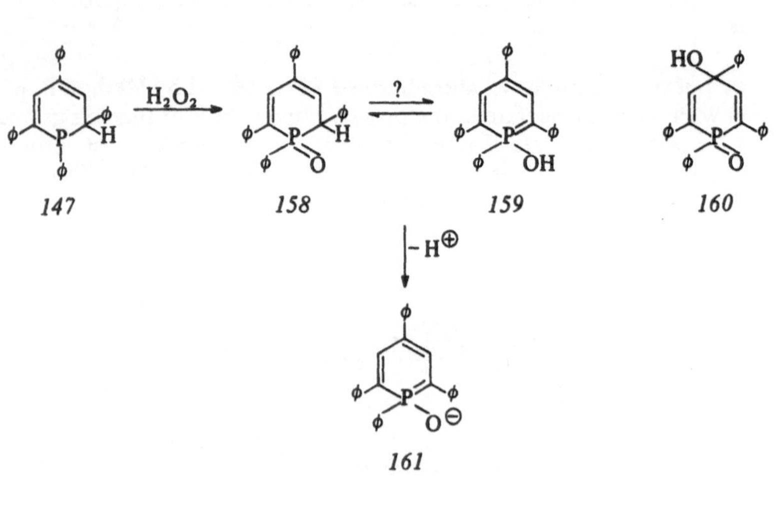

147 *158* *159* *160* *161*

l) Method J: Alkylation of 2-Hydrophosphinic Acids and Esters with Oxonium Salts

According to Teichmann, Jathowski and Hilgetag [89], the P = O group can be alkylated to the $\overset{\oplus}{P}$–O–R group. Similarly, Hettche [88] found that the cyclic 2-hydrophosphinic acids *90* or esters *162* can also be O-alkylated by oxonium salts to form 1,1-dialkoxyphosphorins. This method can also be used to synthesize 1,1-dialkoxyphosphorins having two different alkoxy groups. For example, *162* leads to 1-methoxy-1-ethoxy-2.4.6-triphenyl-λ^5-phosphorin *163*.

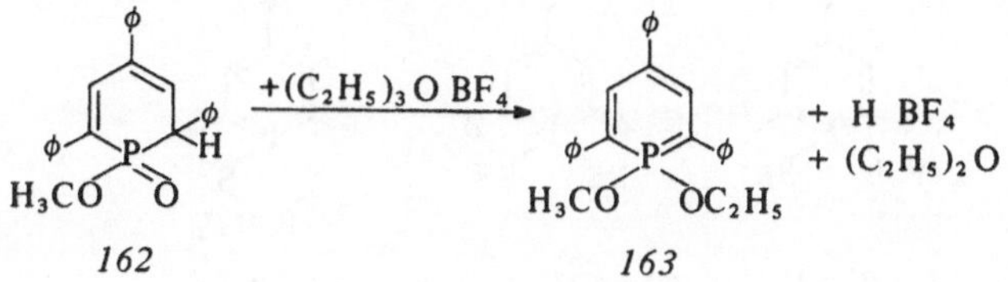

This reaction can also be used to convert the phosphine oxide *158*. (Method I, p. 86) to 1-aryl-1-alkoxy-λ^5-phosphorin *156* [100].

m) Method K: Substitution Reactions of 1.1-Hetero-λ^5-phosphorins at the P Atom

If 1,1-bis-dimethylamino-2.4.6-triphenyl-λ^5-phosphorin *164* is allowed to react with 2 moles of trifluoroacetic acid and a 10-fold molar excess of methanol in refluxing benzene, one of the amino groups is displaced and 1-methoxy-1-dimethylamino-2.4.6-triphenyl-λ^5-phosphorin *167* can be isolated in good yield. The phosphonium salt *165* (H$^\oplus$ can also attack C–2) is initially formed by protonation of *164* at C–4 (or C–2). Nucleophilic substitution then leads to *167* via *166* (Hettche, [88]).

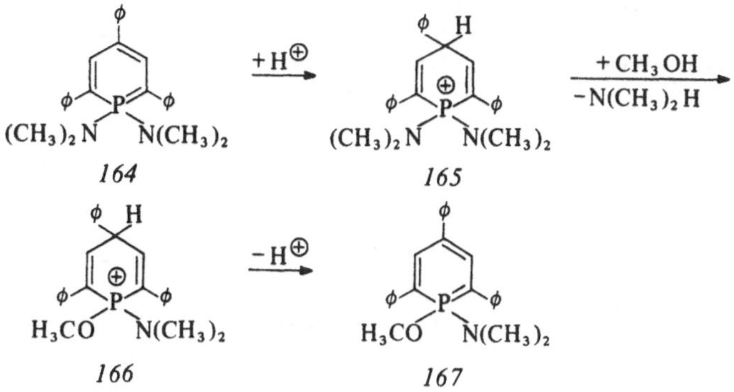

Such nucleophilic substitutions can also be carried out on analogous 1,1-di-arylamino-λ^5-phosphorins and on "oxy-bis"-λ^5-phosphorins *154* or acetates *153*; the latter can be viewed as mixed anhydrides in which phosphinic acid or acetic acid are easily displaced by nucleophiles. These reactions have a broad application in synthesis.

This method was applied by Kanter [92] who succeeded in preparing 1.1-dialkylthio-λ^5-phosphorins *169* by using thiols. These compounds could not been synthesized by any other means.

The intermediate monoalkylthio derivative *168* is rapidly converted to the end product *169*.

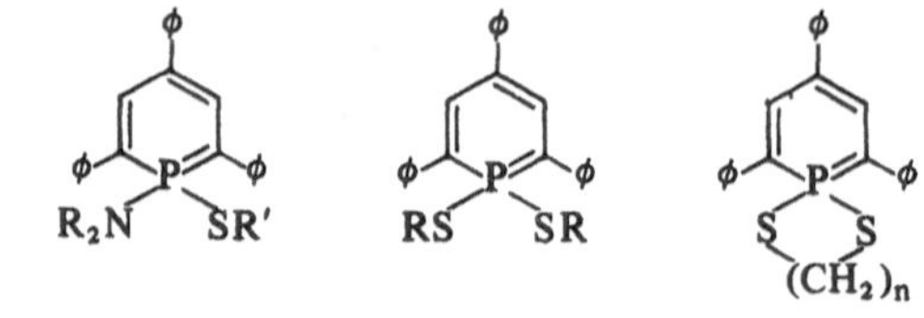

$$(R = R' = C_2H_5) \quad (R = CH_3, C_2H_5) \quad (n = 2,3)$$
$$168 \qquad\qquad 169 \qquad\qquad 170$$

Spiro compounds of the type *170* have also been prepared in this way.

Thiophenol as the nucleophile eliminates the 1.1-substituents to form 2.4.6-triphenyl-λ^3-phosphorin *22*. A similar elimination takes place if 1,1-di-alkyl-thio-λ^5-phosphorins *169* are heated in toluene. This method is thus an excellent way to remove the dialkylamino groups which had been introduced before as protecting groups:

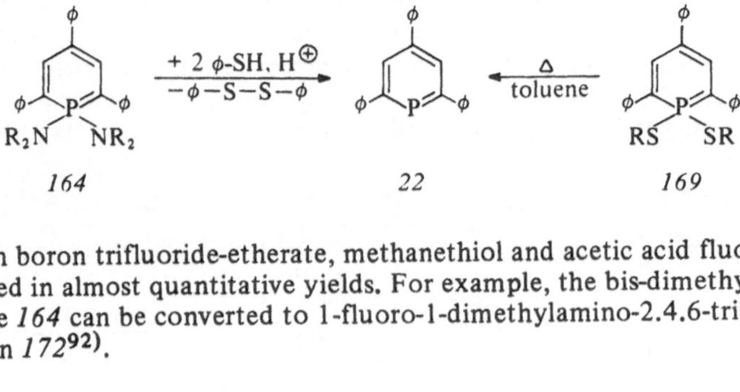

$$164 \qquad\qquad 22 \qquad\qquad 169$$

With boron trifluoride-etherate, methanethiol and acetic acid fluorine is incorporated in almost quantitative yields. For example, the bis-dimethylamino-derivative *164* can be converted to 1-fluoro-1-dimethylamino-2.4.6-triphenyl-phoshorin *172* [92].

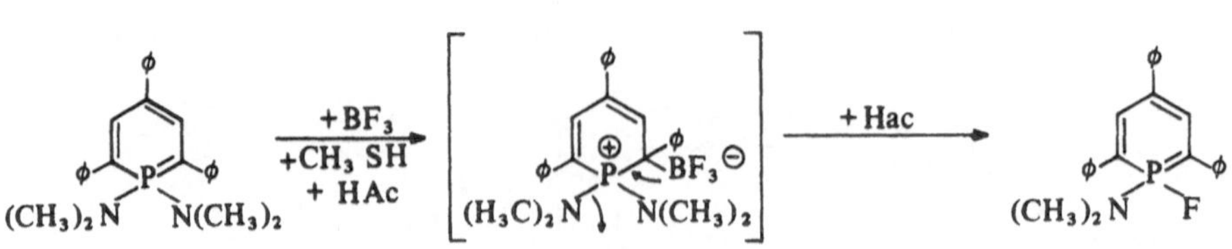

$$164 \qquad\qquad 171 \qquad\qquad 172$$

The mechanism of this reaction remains to be established. It may be that electrophilic addition of BF_3 leads to *171* which then breaks up to form *172*.

According to Kanter[92], the reaction sequence *164* to *169*, if carried out in the presence of trifluoroacetic acid and thiophenol with the careful exclusion of oxygen, leads to a 40% yield of 1-dimethylamino-1-phenoxy-2.4.6-triphenyl-λ^5-

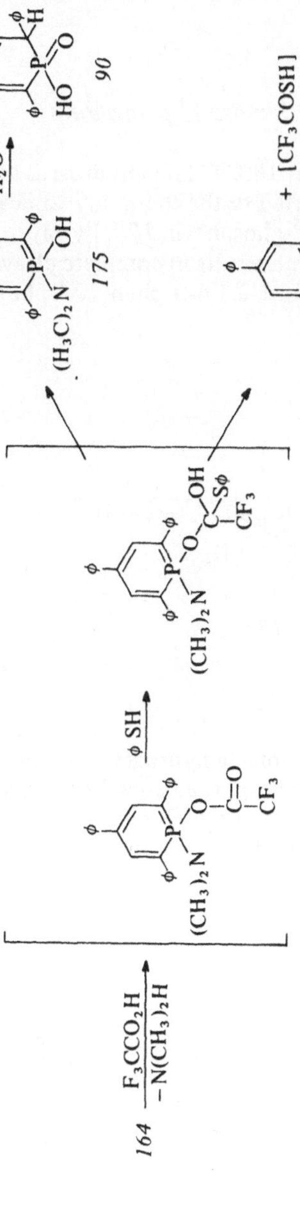

phosphorin *176*. It is conceivable that this reaction proceeds via the trifluoro-acetyl derivative *173* which then adds thiophenol to form *174*. This intermediate can then form the product *176* by migration of the phenyl group, or it can split off the trifluoro acetyl thioester to form *175* which then is hydrolyzed to the 2-hydro-phosphinic acid.

n) Method L: Rearrangements of 1.2-Dihydro-λ^3-phosphorins

Märkl and Merz [87] have shown that at 180 °C 1,2-dihydro-1.2-dibenzyl-2.4.6-triphenyl-λ^3-phosphorin *177* is converted to the thermodynamically more stable isomer 1,1-dibenzyl-2.4.6-triphenyl-λ^5-phosphorin *178*. Pyrolysis at temperatures above 220 °C for longer periods of time results in complete cleavage of both benzyl groups (dibenzyl is formed) to afford 2.4.6-triphenyl-λ^3-phosphorin *22*.

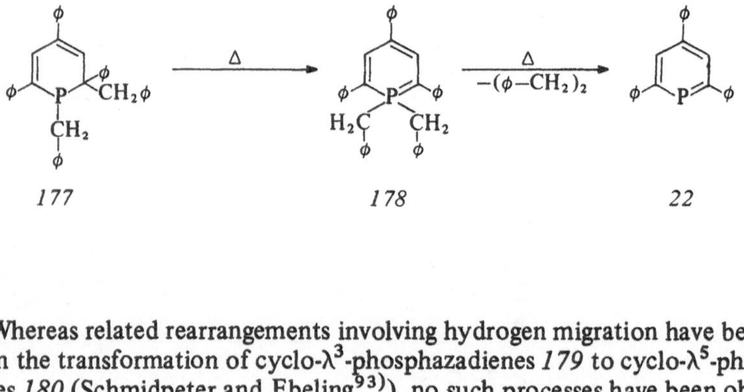

177   *178*   *22*

Whereas related rearrangements involving hydrogen migration have been observed in the transformation of cyclo-λ^3-phosphazadienes *179* to cyclo-λ^5-phosphazatrienes *180* (Schmidpeter and Ebeling[93]), no such processes have been observed for 1,2-dihydro-λ^3-phosphorins, *i. e. 181* fails to rearrange to *182*.

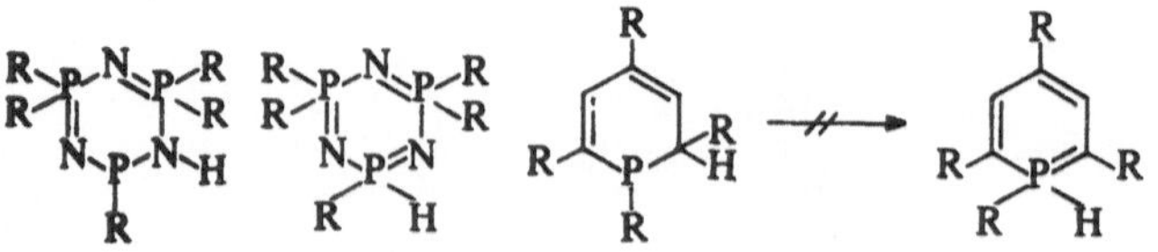

179   *180*   *181*   *182*

Table 20. λ⁵-Phosphorins with 2 OR and with 2 SR residues at the phosphorus

Structure with substituents R^3, R^2, R^1, $R^{1'}$, R^4, R^6 on the phosphorin ring.

R^1	$R^{1'}$	R^2	R^3	R^4	R^6	m.p. °C	Method	Lit.
OCH$_3$	OCH$_3$	C(CH$_3$)$_3$	H	C(CH$_3$)$_3$	C(CH$_3$)$_3$	86	G	44)
OCH$_3$	OCH$_3$	C(CH$_3$)$_3$	H	C$_2$H$_5$	C(CH$_3$)$_3$	146	G	44)
OCH$_3$	OCH$_3$	C(CH$_3$)$_3$	H	CH(CH$_3$)$_3$	C(CH$_3$)$_3$	42–3	G	100)
OCH$_3$	OCH$_3$	C(CH$_3$)$_3$	H	C$_6$H$_4$OCH$_3$	C(CH$_3$)$_3$	119	G	44)
OCH$_3$	OCH$_3$	CH$_3$	H	C$_6$H$_5$	C$_6$H$_5$	104	G	67)
OCH$_3$	OCH$_3$	C$_6$H$_5$	H	CH$_3$	C$_6$H$_5$	141–2	G	43, 67)
OCH$_3$	OCH$_3$	C$_6$H$_5$	H	CH$_2^{\oplus}$ BF$_4^{\ominus}$	C$_6$H$_5$	160	G	67, 96)
OCH$_3$	OCH$_3$	C$_6$H$_5$	H	CH$_2$CH$_3$	C$_6$H$_5$	70	G	43)
OCH$_3$	OCH$_3$	C$_6$H$_5$	H	$^{\oplus}$CH–CH$_3$BF$_4^{\ominus}$	C$_6$H$_5$	134–5	G	96)
OCH$_3$	OCH$_3$	C$_6$H$_5$	H	CH(CH$_3$)$_2$	C$_6$H$_5$	76	G	43, 100)
OCH$_3$	OCH$_3$	C$_6$H$_5$	H	$^{\oplus}$C(CH$_3$)$_2$BF$_4^{\ominus}$	C$_6$H$_5$	157–8	G	43, 96)
OCH$_3$	OCH$_3$	C$_6$H$_5$	H	CH$_2$CN	C$_6$H$_5$	102–3	G	67)
OCH$_3$	OCH$_3$	C$_6$H$_5$	H	CH$_2$SCN	C$_6$H$_5$	140	G	67)
OCH$_3$	OCH$_3$	C$_6$H$_5$	H	CH$_2$-C$_6$H$_5$	C$_6$H$_5$	101	G	67)
OCH$_3$	OCH$_3$	C$_6$H$_5$	H	$^{\oplus}$CH–C$_6$H$_5$BF$_4^{\ominus}$	C$_6$H$_5$	148–9	G	67, 96)
OC$_2$H$_5$	OC$_2$H$_5$	C$_6$H$_5$	H	CH$_2^{\oplus}$ BF$_4^{\ominus}$	C$_6$H$_5$	120–2	G	67, 96)
OC$_2$H$_5$	OC$_2$H$_5$	CH$_2$CH$_2$–C$_6$H$_4$(CH$_3$)		C$_6$H$_5$	C(CH$_3$)$_3$	137–40	G	37)

91

Table 20 (continued)

R^1	$R^{1'}$	R^2	R^3	R^4	R^6	m. p. °C	Method	Lit.
OCH_3	OCH_3	C_6H_5	H	C_6H_5	C_6H_5	112	G	45, 61)
OCD_3	OCD_3	C_6H_5	H	C_6H_5	C_6H_5	110–3	G	45, 61)
OCH_3	OCH_3	C_6D_5	H	C_6D_5	C_6D_5	112	G	45)
OCD_3	OCD_3	C_6D_5	H	C_6D_5	C_6D_5	112	G	45)
OCH_3	OCH_3	C_6H_5	H	C_6H_4Cl	C_6H_5	–	G	45)
OCH_3	OCH_3	C_6H_5	H	$C_6H_4OCH_3$	C_6H_5	106–7	G	45)
OCH_3	OCH_3	$C_6H_4OCH_3$	H	C_6H_5	$C_6H_4OCH_3$	124	G	45)
OCH_3	OCH_3	$C_6H_3[C(CH_3)_3]_2$ 3'5'	H	C_6H_5	$C_6H_3[C(CH_3)_3]_2$ 3'5'	165.5	G	45, 63)
OCD_3	OCD_3	$C_6H_3[C(CH_3)_3]_2$ 3'5'	H	C_6H_5	$C_6H_3[C(CH_3)_3]_2$ 3'5'	165–7	G	45)
OCH_3	OCH_3	$C_6H_4CH_3$	H	$C_6H_4CH_3$	$C_6H_4CH_3$	167	G	45)
OCH_3	OCH_3	$C_6H_4OCH_3$	H	$C_6H_4OCH_3$	$C_6H_4OCH_3$	148–9	G	45, 61)
OC_2H_5	OC_2H_5	C_6H_5	H	C_6H_5	C_6H_5	107	G	45, 61)
OC_2H_5	OC_2H_5	$C_6H_4OCH_3$	H	C_6H_5	$C_6H_4OCH_3$	127	G	45)
$OCH(CH_3)_2$	$OCH(CH_3)_2$	C_6H_5	H	C_6H_5	C_6H_5	112	G	45)
OCH_3	OC_2H_5	C_6H_5	H	C_6H_5	C_6H_5	109–10	J, K	88)
OCH_3	$OCH(CH_3)_2$	C_6H_5	H	C_6H_5	C_6H_5	149–50	J, K	88)
OC_6H_5	OC_6H_5	$C(CH_3)_3$	H	$C(CH_3)_3$	$C(CH_3)_3$	103	G	100, 75)
OC_6H_5	OC_6H_5	$C(CH_3)_3$	H	$C_6H_4OCH_3$	$C(CH_3)_3$	204	G	44)
OC_6H_5	OC_6H_5	C_6H_5	H	CH_3	C_6H_5	175	G	67)

Table 20 (continued)

R^1	$R^{1'}$	R^2	R^3	R^4	R^6	m.p. °C	Method	Lit.
OC_6H_5	OC_6H_5	CH_3	H	C_6H_5	C_6H_5	95–6	G	67)
OC_2H_5	OC_2H_5	C_6H_5	H	CH_3	C_6H_5	95–6	G	67)
OC_2H_5	$OCH(CH_3)_2$	C_6H_5	H	C_6H_5	C_6H_5	85–6	J, K	88)
OCH_2-CCl_3	OCH_2CCl_3	C_6H_5	H	C_6H_5	C_6H_5	160–3	G	45)
OCH_2-CCl_3	OCH_2CCl_3	$C_6H_4OCH_3$	H	C_6H_5	$C_6H_4OCH_3$	157–8	G	45)
OCH_3	OCH_2-CH_2OH	C_6H_5	H	C_6H_5	C_6H_5	Oil	G	45)
OCH_3	$O(CH_2)_3OH$	C_6H_5	H	C_6H_5	C_6H_5	Oil	G	45)
OCH_3	$O(CH_2)_3OCOCH_3$	C_6H_5	H	C_6H_5	C_6H_5	Oil	G	45)
OC_6H_5	OC_6H_5	C_6H_5	H	C_6H_5	C_6H_5	152–4	G	45)
$OC_6H_4OCH_3$	$OC_6H_4OCH_3$	$C_6H_4OCH_3$	H	C_6H_5	$C_6H_4OCH_3$	135–6	G	45)
OC_6H_4Cl	OC_6H_4Cl	$C_6H_4OCH_3$	H	C_6H_5	$C_6H_4OCH_3$	139–41	G	45)
$OC_6H_4NO_2$	$OC_6H_4NO_2$	$C_6H_4OCH_3$	H	C_6H_5	$C_6H_4OCH_3$	192	G	45, 61)
OCH_3	O Tri*)	C_6H_5	H	C_6H_5	C_6H_5	114	E	45)
O Tri*)	O Tri*)	C_6H_5	H	C_6H_5	C_6H_5	247–51	E	45, 99)
O TriOCH₃**)	O TriOCH₃**)	$C_6H_4OCH_3$	H	C_6H_5	$C_6H_4OCH_3$	162	E	45)
OCH_3	$OCOCH_3$	C_6H_5	H	C_6H_5	C_6H_5	127–9	G	88)
OC_2H_5	$OCOCH_3$	C_6H_5	H	C_6H_5	C_6H_5	120–3	G	88)
$OCH(CH_3)_2$	$OCOCH_3$	C_6H_5	H	C_6H_5	C_6H_5	121–2	G	88)
$OC(CH_3)_3$	$OCOCH_3$	C_6H_5	H	C_6H_5	C_6H_5	123–5	G	88)
$OCH(CH_3)_2$	$OCOC_6H_5$	C_6H_5	H	C_6H_5	C_6H_5	160–1	G	101)
$OCH(CH_3)_2$	$OCOC_6H_4CH_3$ (4)	C_6H_5	H	C_6H_5	C_6H_5	153–4	G	101)

93

Table 20 (continued)

R¹	R¹'	R²	R³	R⁴	R⁶	m.p. °C	Method	Lit.
OCH(CH₃)₂	OCOC₆H₄Cl (4)	C₆H₅	H	C₆H₅	C₆H₅	147–9	G	101)
OCH(CH₃)₂	OCOC₆H₄NO₂ (4)	C₆H₅	H	C₆H₅	C₆H₅	148	G	101)
OCOC₆H₅	OCOC₆H₅	C₆H₅	H	C₆H₅	C₆H₅	128–30	G	101)
$-O-CH_2-CH_2-O-$		C(CH₃)₃	H	C₆H₄OCH₃	C(CH₃)₃	162	G	44)
$-O-CH_2-CH_2-O-$		C₆H₅	H	C₆H₅	C₆H₅	179–81	G	45)
$-O-CD_2-CD_2-O-$		C₆H₅	H	C₆H₅	C₆H₅	178	G	45)
$-O-CD_2-CD_2-O-$		C₆D₅	H	C₆D₅	C₆D₅	178	G	45)
$-O-CH_2-CHCH_3-O-$		C₆H₅	H	C₆H₅	C₆H₅	161	G	45)
$-O-C(CH_3)_2-C(CH_3)_2-O-$		C₆H₅	H	C₆H₅	C₆H₅	230–2	G	45)

Table 20 (continued)

R¹	R¹'	R²	R³	R⁴	R⁶	m. p. °C	Method	Lit.
	 ⟨O–benzene–O⟩	$C(CH_3)_3$	H	$C_6H_4OCH_3$	$C(CH_3)_3$	Oil	G	44)
	⟨O–benzene–O⟩	C_6H_5	H	C_6H_5	C_6H_5	183–4	G	45)
	$O(CH_2)_3\text{-}O$	C_6H_5	H	C_6H_5	C_6H_5	197	G	45)
	$O(CH_2)_4\text{-}O$	C_6H_5	H	C_6H_5	C_6H_5	143–6	G	45)
SCH_3	SCH_3	C_6H_5	H	C_6H_5	C_6H_5	146–7	K	92, 95)
SC_2H_5	SC_2H_5	C_6H_5	H	C_6H_5	C_6H_5	104–5	K	92, 95)
	$S\text{–}CH_2\text{–}CH_2\text{–}S$	C_6H_5	H	C_6H_5	C_6H_5	165–70	K	92)
	$S\text{–}(CH_2)_3\text{–}S$	C_6H_5	H	C_6H_5	C_6H_5	150–2	K	92)

*) Tri = (benzene ring with C_6H_5, C_6H_5, C_6H_5)   **) Tri OCH₃ = (benzene ring with C_6H_5, C_6H_5, $C_6H_4OCH_3(4)$)

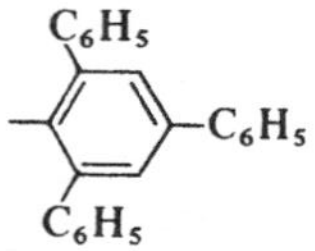

95

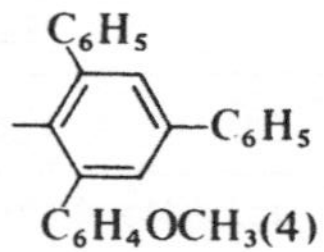

Table 21. λ^5-Phosphorins with 2 NR_2-residues at the phosphorus

R^1	$R^{1'}$	R^2	R^4	R^6	m. p. °C	Method	Lit.
$N(CH_3)_2$	$N(CH_3)_2$	CH_3	C_6H_5	C^6H^5	141	G	67)
$N(CH_3)_2$	$N(CH_3)_2$	C_6H_5	CH_3	C_6H_5	122–3	G	67)
$N(CH_3)_2$	$N(CH_3)_2$	C_6H_5	$CH_2\overset{\oplus}{}BF_4^{\ominus}$	C_6H_5	135	G	67)
$N(CH_3)_2$	$N(CH_3)_2$	C_6H_5	C_6H_5	C_6H_5	121–2	G	88, 94)
$N(C_2H_5)_2$	$N(C_2H_5)_2$	C_6H_5	C_6H_5	C_6H_5	126–7	G	88, 94)
$N[CH(CH_3)_2]H$	$N[CH(CH_3)_2]H$	C_6H_5	C_6H_5	C_6H_5	181–2	G	88, 94)
$HN-CH_2-CH_2-NH$		C_6H_5	C_6H_5	C_6H_5	193–7	G	88, 94)
$H_3CN-CH_2-CH_2-CH_2-NCH_3$		C_6H_5	C_6H_5	C_6H_5	106–7	G	88, 94)
$N(CH_3)_2$	$N(CH_3)_2$	$C_6H_3[C(CH_3)_3]_2$	C_6H_5	$C_6H_3[C(CH_3)_3]_2$	229–3	G	63)
$N(C_6H_5)_2$	$N(C_6H_5)_2$	$C_6H_4CH_3$	C_6H_5	C_6H_5	179	G	88, 94, 99)
$N(C_6H_4CH_3)_2$	$N(C_6H_4CH_3)_2$	C_6H_5	C_6H_5	C_6H_5	168–70	G	88, 94)
$N(C_6H_4CH_3)_2$	$N(C_6H_4CH_3)_2$	C_6H_5	$C_6H_4OCH_3$	C_6H_5	168–9	G	88, 94)
$N(CH_3)_2$	$N(CH_3)_2$	C_6H_5	C_6H_5	C_6H_5	185–7	G	88, 94)
$N(CH_3)_2$	$N(CH_3)_2$	C_6H_5	CH_3	C_6H_5	122–3	G	67)
$N(CH_3)_2$	$N(CH_3)_2$	C_6H_5	$CH_2\overset{\oplus}{}BF_4^{\ominus}$	C_6H_5	133–5	G	67, 96)

Table 22. λ^5-Phosphorins with one -OR and one -NR$_2$ residue at the phosphorus

R^1	R$^{1'}$	R^2	R^4	R^6	m. p. °C	Method	Lit.
N(CH$_3$)$_2$	OCH$_3$	C$_6$H$_5$	C$_6$H$_5$	C$_6$H$_5$	140	K	88, 95)
N(CH$_3$)$_2$	OC$_2$H$_5$	C$_6$H$_5$	C$_6$H$_5$	C$_6$H$_5$	99–100	K	88, 95)
N(C$_2$H$_5$)$_2$	OCH$_3$	C$_6$H$_5$	C$_6$H$_5$	C$_6$H$_5$	83–6	K	88, 95)
N(C$_2$H$_5$)$_2$	OC$_2$H$_5$	C$_6$H$_5$	C$_6$H$_5$	C$_6$H$_5$	98–100	K	88, 95)
NHCH(CH$_3$)$_2$	OCH$_3$	C$_6$H$_5$	C$_6$H$_5$	C$_6$H$_5$	155–6	K	88)
N(C$_2$H$_5$)$_2$	OC$_6$H$_5$	C$_6$H$_5$	C$_6$H$_5$	C$_6$H$_5$	114–5	–	92)
N(C$_2$H$_5$)$_2$- H$_3$CN-CH$_2$-CH$_2$-O		C$_6$H$_5$	C$_6$H$_5$	C$_6$H$_5$	149–51	G	88, 94)
N(C$_6$H$_5$)$_2$	OCH$_3$	C$_6$H$_5$	C$_6$H$_5$	C$_6$H$_5$	192–3	K, F	88)
N(C$_6$H$_5$)$_2$	OC$_2$H$_5$	C$_6$H$_5$	C$_6$H$_5$	C$_6$H$_5$	143–5	K, F	88, 95)
N(C$_6$H$_5$)$_2$	OCH(CH$_3$)$_2$	C$_6$H$_5$	C$_6$H$_5$	C$_6$H$_5$	165–6	K	88)
N(C$_6$H$_5$)$_2$	O(C$_6$H$_4$CH$_3$)	C$_6$H$_5$	C$_6$H$_5$	C$_6$H$_5$	–	K	88)
N(C$_6$H$_4$CH$_3$)$_2$	OCH$_3$	C$_6$H$_5$	C$_6$H$_5$	C$_6$H$_5$	195	K, F	88, 95)
N(C$_6$H$_4$CH$_3$)$_2$	OC$_2$H$_5$	C$_6$H$_5$	C$_6$H$_5$	C$_6$H$_5$	171–2	F, K	88)
N(C$_6$H$_4$CH$_3$)$_2$	OCH(CH$_3$)$_2$	C$_6$H$_5$	C$_6$H$_5$	C$_6$H$_5$	194–5	K, F	88)
N(C$_6$H$_4$CH$_3$)$_2$	OCH$_2$-CH$_3$OH	C$_6$H$_5$	C$_6$H$_5$	C$_6$H$_5$	161–2	K	88)
N(C$_6$H$_4$CH$_3$)$_2$	OCH$_2$-C$_6$H$_5$	C$_6$H$_5$	C$_6$H$_5$	C$_6$H$_5$	143–5	K	88)
N(C$_2$H$_5$)$_2$	OCOCH$_3$	C$_6$H$_5$	C$_6$H$_5$	C$_6$H$_5$	108–10	G	88)

Table 23. λ^5-Phosphorins with one C- and one O- or S-residue at

R^1	$R^{1'}$	R^2	R^3	R^4	R^6	m. p. °C	λ_{max_1}
CH_3	OCH_3	C_6H_5	H	C_6H_5	C_6H_5	137–9	426
CH_3	OC_2H_5	C_6H_5	H	C_6H_5	C_6H_5	108–9	426
CH_3	$OCH(CH_3)_2$	C_6H_5	H	C_6H_5	C_6H_5	156	433
CH_3	$OC(CH_3)_3$	C_6H_5	H	C_6H_5	C_6H_5	137–41	434
CH_2CH_3	OC_2H_5	C_6H_5	H	C_6H_5	C_6H_5	137–8	426
CH_2CH_3	$OCH(CH_3)_2$	C_6H_5	H	C_6H_5	C_6H_5	134–5	–
CH_2CH_3	$OC(CH_3)_3$	C_6H_5	H	C_6H_5	C_6H_5	129–31	432
$CH_2C_6H_5$	OCH_3	C_6H_5	H	C_6H_5	C_6H_5	133	426
$CH_2C_6H_5$	$OCH_2C_6H_5$	C_6H_5	H	C_6H_5	C_6H_5	118–9	427
C_6H_5	OC_2H_5	CH_3	H	C_6H_5	C_6H_5	70–1	409
C_6H_5	OCH_3	C_6H_5	H	C_6H_5	C_6H_5	136–7	426
C_6H_5	OC_2H_5	C_6H_5	H	C_6H_5	C_6H_5	126–7	–
C_6H_5	$OCH(CH_3)_2$	C_6H_5	H	C_6H_5	C_6H_5	160–1	434
C_6H_5	$OCH_2C_6H_5$	C_6H_5	H	C_6H_5	C_6H_5	154–5	–
C_6H_5	OC_2H_5	$-(CH_2)_4-$		C_6H_5	C_6H_5	142–3	402
C_6H_5	OC_6H_5	C_6H_5	H	C_6H_5	C_6H_5	186–7	419
C_6H_5	O-Tri **)	C_6H_5	H	C_6H_5	C_6H_5	213–5	428
C_6H_5	SC_2H_5	C_6H_5	H	C_6H_5	C_6H_5	82–4	436

*) See also Table 14, p. 66.
**) Tri = 2.4.6-triphenylphenyl.

Ar = C_6H_5

Table 24. 1-Alkoxy or 1-dialkylamino-1-oxy-bis-2.4.6-triphenyl-λ^5-phosphorins

R^1	$R^{1'}$	m. p. °C	Method	Lit.
OCH_3	OCH_3	152–4	G	88, 94)
OC_2H_5	OC_2H_5	120–2	G	88, 94)

the phosphorus *)

ϵ_1	λ_{max2}	ϵ_2	λ_{max3}	ϵ_3	Method	Lit.
17500	313	17500	273	15800	K	88, 100)
20500	316	18000	275	19900	K	88)
21400	–	–	279	33000	K	88)
22200	320	18200	282	17600	K	88)
17800	318	18400	272	14700	K	88)
–	–	–	–	–	K	88)
21700	321	19400	279	17300	K	88)
18700	317	19720	273	17640	I	90)
15340	318	16640	274	16320	I	90)
8000	317	13600	264	10250	I	90)
11400	314	14050	272	14400	I, H	90)
–	–	–	–	–	I	90)
20400	320	20300	285	17600	K	88)
–	–	–	–	–	I	90)
1200	309	9000	264	11400	I	90)
11500	310	14000	274	16000	I, H	90)
10700	331	15500	–	–	H	86)
–	–	–	–	–	I	90)

Table 24 (continued)

R^1	$R^{1'}$	m. p. °C	Method	Lit.
$OCH(CH_3)_2$	$OCH(CH_3)_2$	161–3	G	88, 94)
$OC(CH_3)_3$	$OC(CH_3)_3$	129–30	G	88, 94)
OCH_3	$OCH(CH_3)_2$	144–6	G	88, 94)
$NHCH(CH_3)_2$	$NHCH(CH_3)_2$	129–32	G	88)
$N(CH_3)_2$	$N(CH_3)_2$	174–7	G	88, 94)
$N(C_2H_5)_2$	$N(C_2H_5)_2$	163–5	G	88, 94)
OCH_3	$N(CH_3)_2$	156–9	G	88, 94)

C. Physical Properties

1. UV and Visible Spectra

In going from λ^3-phosphorins to λ^5-phosphorins, profound changes in the absorption spectra are observed: new bands appear in the visible region. Thus, all λ^5-phosphorins are colored, ranging from yellow to red. Many also show fluorescence.

1.1-Carbo-λ^5-phosphorins having C groups at the phosphorus have the longest wave absorptions. The influence of the particular structure of these substituents is rather small. Thus, 1.1-diphenyl-2.4.6-triphenyl-λ^5-phosphorin and 1.1-dibenzyl-2.4.6-triphenyl-λ^5-phosphorin have very similar absorption spectra.

The steric influence of the P substituents appears to play a more pronounced role, as can be seen by comparing *144*, and *148* (Table 25); the spiro compound absorbs at much shorter wavelength.

The bathochromic influence of substituents at positions 2,4 and 6 is quite distinct in both-λ^5-phosphorin and λ^3-phosphorin compounds (e. g. *118* vs. *144*; *47* vs. *22*, Table 25 and Table 26).

Table 25. Absorption maxima of some 1.1-carbo-λ^5-phosphorins compared with λ^3-phosphorins

118 yellow		*47* colourless		*144* red		*148* orange		*22* colourless	
λ_{max}	ϵ	λ_{max}	ϵ	λ_{max}	ϵ	λ_{max}	ϵ	λ_{max}	ϵ
409	–	–	–	515	8900	474	6300	–	–
–	–	246	8500	342	17000	336	11300	314	12600
–	–	213	19000	–	–	–	–	278	41000

1,1-Carbo-phosphorins are relatively basic. They are protonated by aqueous hydrogen chloride, thereby losing their color. According to Märkl, the UV spectrum of the salt is similar to that of 1,2-dihydro-λ^3-phosphorins (Fig. 25). However, it cannot be concluded on the basis of the UV spectrum alone that protonation takes place at position C–2. The C–4 position can also be protonated, as has been shown by NMR spectroscopy to be the case for 1,1-dimethoxy-2.4.6-triphenyl-λ^5-phosphorin. The influence of acid addition on the UV spectra is similar in both classes of compounds (compare Fig. 25 with Fig. 26). However, it should be noted that aqueous acids cannot be used to protonate 1,1-dimethoxy-2.4.6-triphenyl-λ^5-phos-

Table 26: Influence of the substituents in 2.4.6-position to the absorption spectra of 1.1-dimethoxy-λ^5-phosphorins

	λ_{max_1} (ϵ_1)	λ_{max_2} (ϵ_2)	λ_{max_3} (ϵ_3)	λ_{max_4} (ϵ_4)
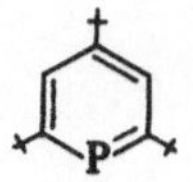	262 (4100)	– –	– –	– –
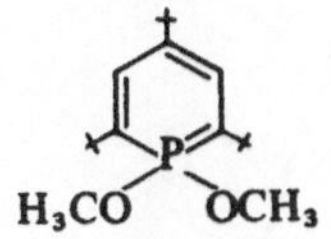 H_3CO OCH_3	354 (10000) *)	262 (2600)	– –	– –
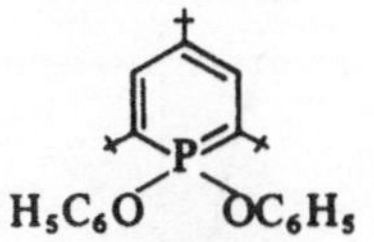 H_5C_6O OC_6H_5	362 (10000) *)	262 (3500)	– –	– –
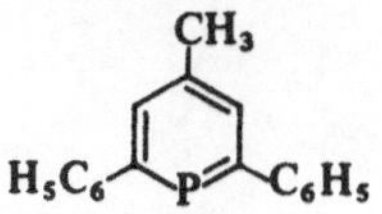 CH_3 H_5C_6 C_6H_5	271 (28100)	– –	– –	– –
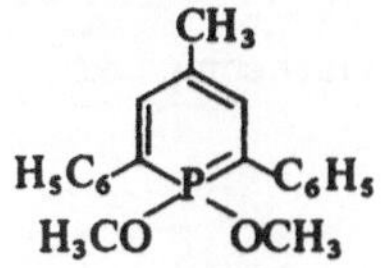 CH_3 H_5C_6 C_6H_5 H_3CO OCH_3	416 (15850)	278 (13500)	238 (19800)	–
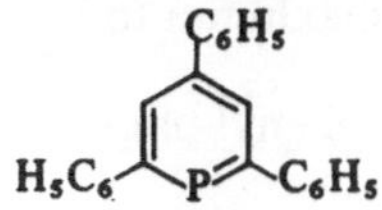 C_6H_5 H_5C_6 C_6H_5	314 (12600)	278 (41000)	– –	– –
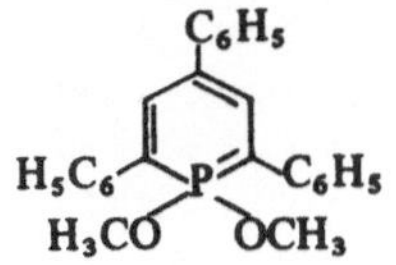 C_6H_5 H_5C_6 C_6H_5 H_3CO OCH_3	417 (13700)	305 (11900)	278 (12700)	220 (16600)

*) In CCl_4

phorin. Upon treatment with base, the original spectra of 1,1-carbophosphorins can be observed; of course, this holds only if oxygen is carefully excluded and if the compounds themselves do not undergo other reactions with base.

More extensive studies have been made on 1,1-hetero-λ^5-phosphorins, especially on those in which the hetero atoms are oxygen [45]. Here again a sizable bathochromic shift in the absorption spectra is observed in going from λ^3-phosphorins to λ^5-phosphorins. Replacement of aliphatic groups by aromatic substituents in the

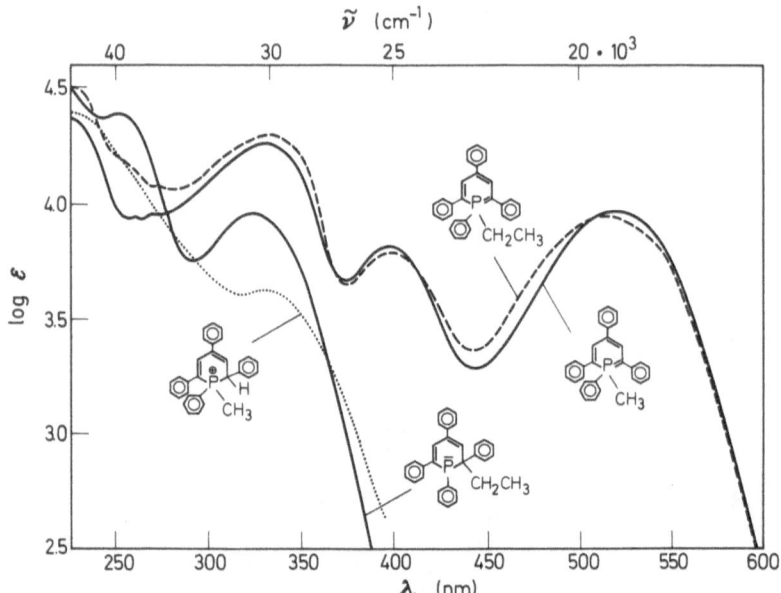

Fig. 25. UV spectra of 1.1-carbo-λ^5-phosphorins and their protonation product compared with a 1.2-substituted λ^3-phosphorin according to Märkl [85]

2.4 and 6 positions of the phosphorin ring leads to a shift of the maximum to longer waves (Table 26).

Substituents in the phenyl groups of 1.1-dimethoxy-2.4.6-tri-phenyl-phosphorin have a small effect on the absorption spectra (Table 27).

Table 27. Influence of the substituents in the phenyl-residues on the absorption spectrum of 1.1-dimethoxy-2.4.6-triphenylphosphorins *)

	λ_{max1}	ϵ_1	λ_{max2}	ϵ_2
$R^2 = R^4 = C_6H_5; R^4 = C_6H_4OCH_3$	425	(25000)	278	(26100)
$R^2 = R^4 = C_6H_5; R^4 = C_6H_4Cl$	417	(17000)	278	(13400)
$R^2 = R^4 = C_6H_4OCH_3; R^4 = C_6H_5$	414	(20800)	288	(19600)
$R^2 = R^4 = R^6 = C_6H_4CH_3$	420	(20300)	286	(21400)
$R^2 = R^4 = R^6 = C_6H_4OCH_3$	417	(22300)	284	(16800)

*) In cyclohexane; all substituents in the position 4′ of the aromatic rings.

In contrast, the nature of the hetero atoms R^1 and $R^{1'}$ in the 2.4.6-triphenyl-λ^5-phosphorins have a very definite effect. Electronic as well as steric factors appear to play a role (see Tables 28 and 29). Electron-withdrawing groups shift the maxima to shorter wavelengths. Bulky substituents cause a shift to longer wavelengths. 1.1-Spiro compounds absorb at extremely short wavelength: the smaller the ring, the shorter the wavelength.

The ring size of the spiro derivatives on the one hand and the bulkiness of the open chain substituents on the other, are expected to have an opposite influence on the size of the O−P−O angle and thereby on the state of hybridization of phosphorus. These geometric factors thus influence the efficiency of phosphorus d-orbital participation. According to Schweig, maximum d_{yz} overlap occurs if the d orbitals are oriented at an angle of 45° with respect to the xy-plane of the ring.

Table 28. Absorptionmaxima of 1.1-hetero-2.4.6-triphenyl-λ^5-phosphorins (in cyclohexane, λ_{max} in nm)

R^1	$R^{1'}$	λ_{max1}	ϵ_1	λ_{max2}	ϵ_2	λ_{max3}	ϵ_3
OCH_3	OCH_3	417	13700	305	11900	278	12700
OCH_3	OC_2H_5	422	21400	310	18400	283	18800
OCH_3	$O(CH_2)_2OH$	420	23100	305	13800	278	24100
OCH_3	$O(CH_2)_2OH$	409	13000	320	3600[b]	245	15400
OCH_3	$O(CH_2)_3OH$	416	6700	–	–	272	12200
OC_2H_5	OC_2H_5	423	20000	310	17600	283	16700
$OCH(CH_3)_2$	$OCH(CH_3)_2$	431	20300	318	15300	285	17900
OCH_2CCl_3	OCH_2CCl_3	411	11800	–	–	281	16100
OC_6H_5	OC_6H_5	411	12300	290	16500	247	12500
$OTri$	$OTri$	425	13700	323	17000	251	9200
*)OC_6H_5	OC_6H_5	410	4400	291	5900	247	4600[a]
*)$OC_6H_4NO_2$	OC_6H_4Cl	411	17000	286	25000	–	–
*)$OC_6H_4NO_2$	$OC_6H_4NO_2$	407	17600	286	45900	286	45900
OCH_3	$N(CH_3)_2$	417	19700	317	19700	273	16200
OCH_3	$N(C_2H_5)_2$	412	21600	315	22600	268	15700
OC_2H_5	$N(CH_3)_2$	421	20700	318	20000	275	16400

103

Table 28 (continued)

R^1	$R^{1\prime}$	λ_{max1}	ϵ_1	λ_{max2}	ϵ_2	λ_{max3}	ϵ_3
OC_2H_5	$N(C_2H_5)_2$	420	15800	320	16600	273	12700
OC_6H_5	$N(C_2H_5)_2$	406	17000	310	19800	267	21300
$N(CH_3)_2$	$N(CH_3)_2$	431	19900	328	16700	267	12800
$N(C_2H_5)_2$	$N(C_2H_5)_2$	458	23800	334	23500	282	16600
$N[CH(CH_3)_2]H$	$N[CH(CH_3)_2]H$	436	16600	327	17800	273	13200
$N(C_6H_5)_2$	$N(C_6H_5)_2$	439	8450	330	19000	–	–
$N(C_6H_4CH_3)_2$	$N(C_6H_4CH_3)_2$	437	9300	333	20700	–	–
SCH_3	SCH_3	438	11600	328	17200	–	–
SC_2H_5	SC_2H_5	432	12700	323	20500	–	–
$N(C_2H_5)_2$	$S(CH_2)_3SH$	427	1500	328	3100[b]	–	–
$N(CH_3)_2$	F	390	20000[c]	–	–	–	–
$N(C_2H_5)_2$	F	390	16000[c]	–	–	–	–
F	F	377	18800	–	–	–	–

*) Derived from 4-phenyl-2.6-(p-methoxyphenyl)-phosphorin.
a) In CH_2Cl_2.
b) In ethanol.
c) In acetone.

Whether a small O–P–O angle (as in the spiro derivatives) raises the energy of the LUMO or lowers the energy of the HOMO (or some of both) cannot be concluded on the basis of present experimental information.

The absorption maxima of "mixed" 1-carbo-1-hetero-2.4.6-tri-phenyl-λ^5-phosphorins having R^1 = alkyl or aryl and $R^{1\prime}$ = O-alkyl, O-aryl, or S-alkyl are given in Table 23 (p. 98). The electronic and steric effects of the substituents are similar to those previously discussed.

The protonation of 1,1-diethoxy-2.4.6-triphenyl-λ^5-phosphorin has been studied in detail [45] and serves as an example for the protonation of 1,1-hetero-λ^5-phosphor-

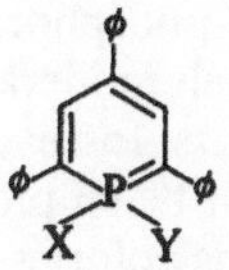

Table 29. Absorptionsmaxima of 1.1-hetero-spiro-2.4.6-triphenyl-λ^5-phosphorins (in cyclohexane, λ_{max} in nm)

X Y	λ_{max_1}	ϵ_1	λ_{max_2}	ϵ_2	λ_{max_3}	ϵ_3
O O CH_2-CH_2	378	19900	298	22700	252	19700
O O CH_2-$CHCH_3$	377	20600	300	16500	–	–
O O $C(CH_3)_2$-$C(CH_3)_2$	357	12000	298	17900	–	–
O O CH_2 CH_2 $\diagdown CH_2 \diagup$	378	21100	296	20100	–	–
O O CH_2 CH_2 CH_2 — CH_2	379	17100	312	19100	250	28800
	381	19000	281	26100	266	28300
O NCH_3 CH_2 – CH_2	389	13500	307	16200	258	13300
CH_3N NCH_3 CH_2-CH_2	407	20700	316	23800	269	18100
HN NH CH_2-CH_2	402	5000	315	6100	260	5000
S S CH_2- CH_2	424	11600	319	6200[a]	–	–
S S CH_2 CH_2 $\diagdown CH_2 \diagup$	437	5700	328	19400[a]	–	–

[a] In ethanol.

105

ins in general. In contrast to 1,1-carbo-λ^5-phosphorins, the hetero-λ^5-phosphorins are not protonated by aqueous HCl or such moderately strong organic acids as dichloroacetic acid in non-aqueous solvents. Instead, trifluoroacetic acid in nonaqueous solvents such as cyclohexane must be employed to completely protonate the λ^5-phosphorin. Under these conditions colorless λ^5-phosphorin salts are indeed formed (Fig. 26). The addition of K-tert-butoxid in tert-butanol causes deprotonation; the spectrum indicates almost quantitative reconversion to the starting material. The NMR spectrum (p. 117) shows that protonation takes place at positions C—2 and C—4 in a ratio of 3:1. Precise measurements of the basicity of λ^5-phosphorins have not yet been carried out.

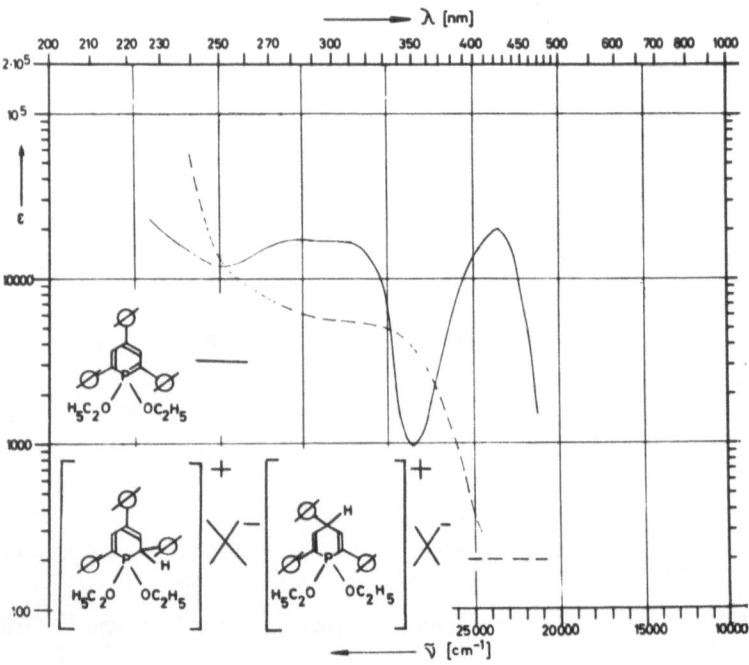

Fig. 26. UV spectra of 1.1-diethoxy-2.4.6-triphenyl-λ^5-phosphorin in cyclohexane and in trifluoroacetic acid

2. Fluorescence and Fluorescence Spectra

Most of the light yellow to yellowish-green colored 2.4.6-aryl-substituted λ^5-phosphorins fluoresce strongly. This phenomenon is an excellent aid in tracing λ^5-phosphorins in preparative work. 1.1-Bis-[4-nitro-phenoxy-] 2.4.6-triphenyl-λ^5-phosphorin and the 1.1-dialkylthio-2.4.6-triphenyl-λ^5-phosphorins do not fluorescence.

The fluorescence spectra of a large number of these compounds have been measured [102])*

An example is given in Fig. 27.

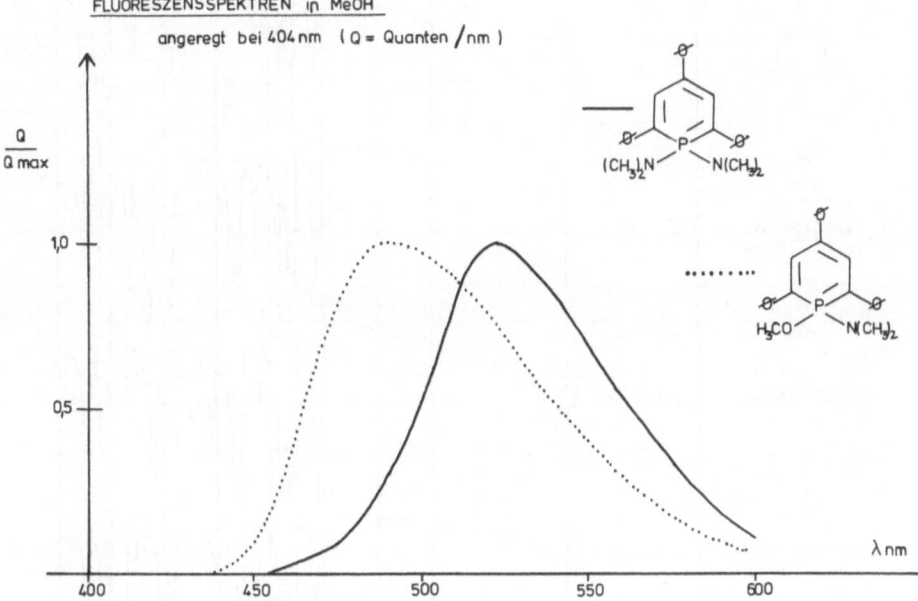

Fig. 27. Fluorescence spectra of 1-dimethylamino-1-methoxy- and of 1.1-bis-dimethylamino-2.4.6-triphenyl-λ^5-phosphorins in methanol

3. I R Spectra

λ^3-Phosphorins as well as λ^5-phosphorins have five characteristic bands between 1400 and 1600 cm^{-1}. 1.1-Dialkoxy- and 1.1-diaryloxy-λ^5-phosphorins have additional intense bands in the region 1180–1220 cm^{-1} and at 1008 cm^{-1} and 1040 cm^{-1}, as well as a weak band at 1160 cm^{-1} which can be attributed to the P–O vibration. In P–N compounds the band at 718 cm^{-1} is probably due to the P–N vibration. With increasing P–N band strength it shifts to 785 cm^{-1} Fig. 28 reviews the results.

*) We have to thank Prof. Dr. F. P. Schäfer, M. P. I. für Biophysikalische Chemie, Göttingen, for the data of the fluorescence spectra.

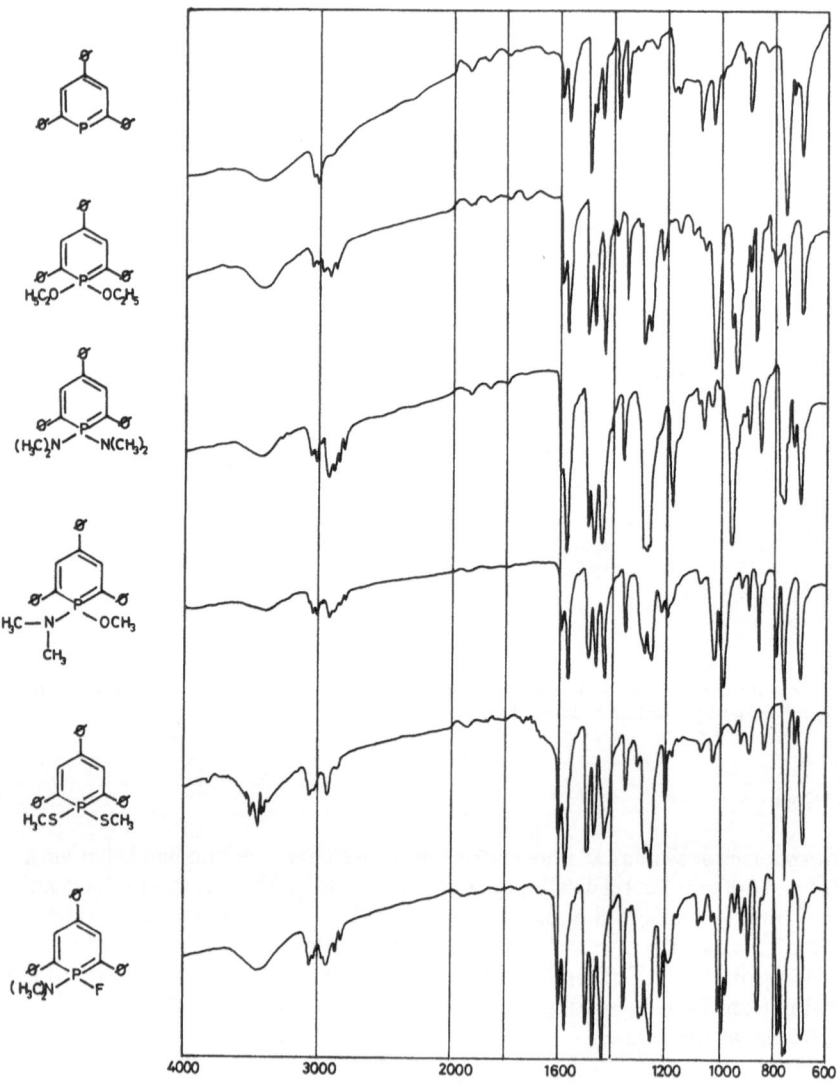

Fig. 28. IR spectra of λ^3-phosphorin and λ^5-phosphorins in KBr

4. NMR Spectra

The ^{1}H-NMR spectra of all 1.1-carbo or 1.1-hetero-λ^5-phosphorins show a doublet between $\delta = 7,5$ and $\delta = 8,5$ ppm with $J_{P-C-C-H} = 30$ to 50 Hz which is due to the protons at C−3 and C−5 of the phosphorin ring. The low-field signals usually appear somewhat lower than those of λ^3-phosphorins. The position of these signals suggests the existence of a ring current induced by the aromatic λ^5-phosphorin system. However, the vinyl protons of λ^5-phospha-cyclohexadiene -2,5 or -2,4 derivatives also absorb at relative low fields (p. 135). Much more characteristic are the P−H coupling constants, which are about six times as large in λ^5-phosphorins as in λ^3-phosphorins ($J_{P-C-C-H} = 5-7$ Hz). Indeed, they provide an excellent help in the identification of λ^5-phosphorins.

By comparing the spectra of 1.1-dimethoxy-2.4.6-triphenyl-λ^5-phosphorin and 1.1-dimethoxy-2.4.6-tris-(pentadeuterophenyl)-λ^5-phosphorin (Figs. 29 and 30), the coupling of the meta protons can easily be singled out [45]. Compare with Figs.

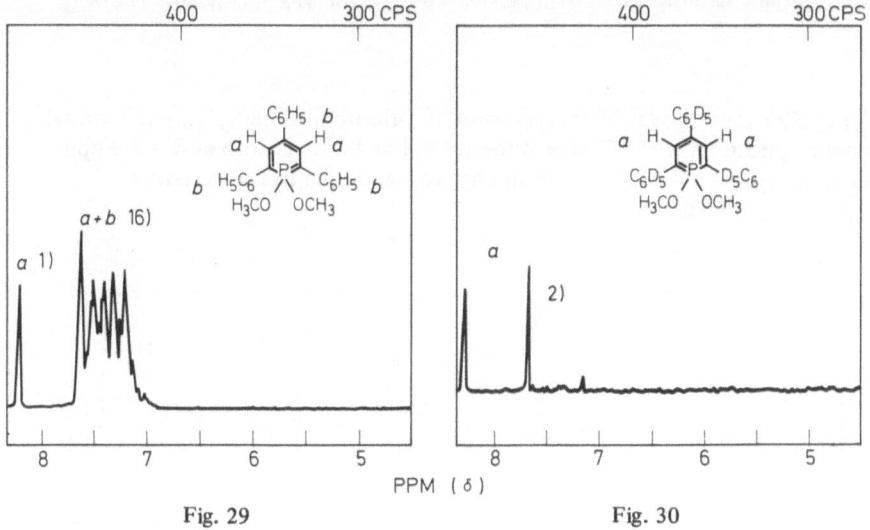

Fig. 29   Fig. 30

Figs. 29 and 30. ^{1}H-NMR spectrum of 1.1-dimethoxy-2.4.6-triphenyl-λ^5-phosphorin and of 1.1-dimethoxy-2.4.6-tris-pentadeuterophenyl-λ^5-phosphorin in $CDCl_3$

9 and 10 p. 32. In the 1.1-bis-tri-deuteromethoxy-derivative the signals of the equivalent methoxy groups at $\delta = 3,36$ ppm $J_{P-C-H} = 13,8$ Hz also disappear.

The ^{1}H-NMR spectrum of 1.1-dimethoxy-2.4.6-tri-tert-butyl-λ^5-phosphorin (Fig. 31), is in principle very similar to that of 1.1-bis-diethylamino-2.4.6-tri-phenyl-λ^3-phosphorin, or 1.1-dimethylthio-2.4.6-tri-phenyl-λ^5-phosphorin. The changes in the ^{1}H-NMR spectra which accompany protonation of 1.1-hetero-λ^5-phosphorins are discussed on p. 117.

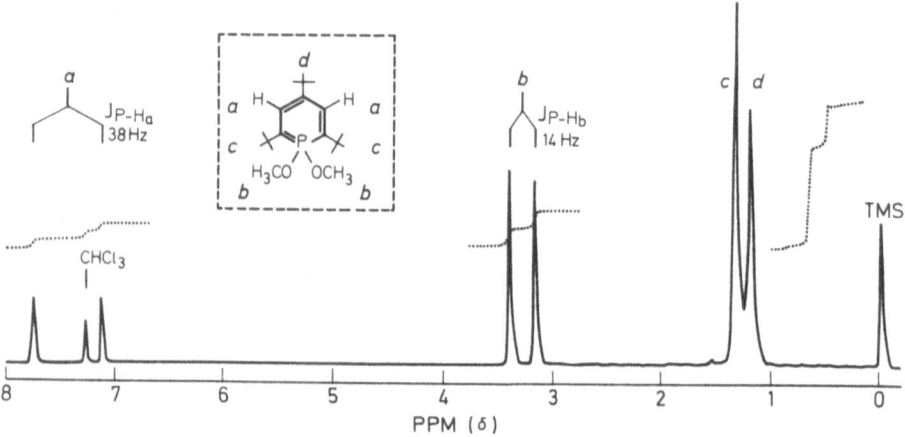

Fig. 31. ^{1}H-NMR spectrum of 1.1-dimethoxy-2.4.6-tri-tert-butyl-λ^5-phosphorin in CDCl$_3$

Fig. 32 shows the ^{1}H-NMR spectrum of 1-fluoro-1-dimethylamino-2.4.6-tri-phenyl-λ^5-phosphorin [92]. The protons at C–3 and C–5 absorb at $\delta = 7,9$ ppm with $J_{P-C-C-H} = 37$ Hz; fluorine and hydrogen couple to the extent of $J_{F-P-C-C-H} = 6$ Hz.

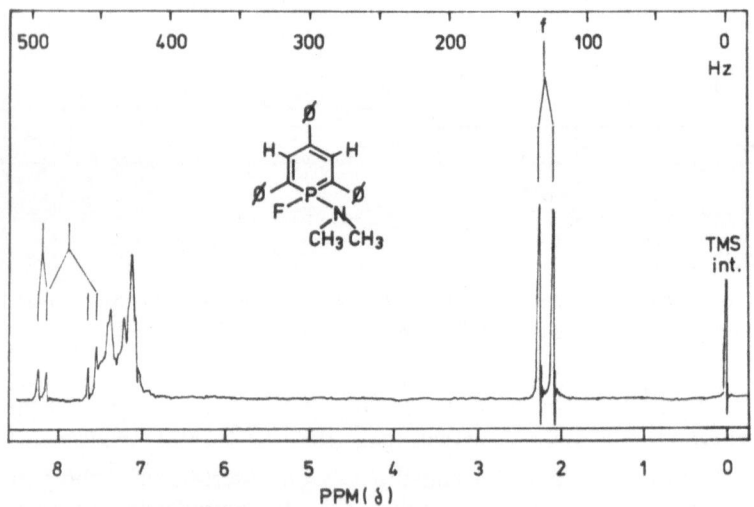

Fig. 32. ^{1}H-NMR spectrum of 1-dimethylamino-1-fluoro-2.4.6-triphenyl-λ^5-phosphorin in C$_6$D$_6$

The methyl protons of the dimethylamino group appear at $\delta = 2.17$ ppm with $J_{P-N-C-H} = 10$ Hz. The phosphorus fluorine coupling has a value of $J_{P-F} = 1035$ Hz; the ^{19}F-signal of the diethylamino compound comes at $\delta = +47.02$ ($J_{P-F} = 1020$ Hz), that of ^{31}P at $\delta = -58.3$ ppm (see Table 30).

The ^{31}P spectra of 1.1-carbo- and 1.1-hetero-λ^5-phosphorins are quite revealing (Table 30). The chemical shift of the ^{31}P signals depends closely upon the electronegativity of the 1.1-substituents. A relationship beetween the ^{31}P-chemical shift and the long-wave absorption has also been established (see p. 103).

Table 30. ^{31}P-NMR-signals (δ in ppm, 85% H_3PO_4 as external standard) of 2.4.6-triphenyl-λ^5-phosphorins

R^1	$R^{1'}$	(ppm)	Solvent	Lit.
2.4.6-triphenyl-λ^3-phosphorin*)		-178.2	Benzene	9)
OCH_3	OCH_3	$- 65.2$	Benzene	45)
OC_2H_5	OC_2H_5	$- 59.3$	Benzene	45)
F	$N(CH_3)_2$	$- 58.3$	Benzene	92)
$OCH(CH_3)_2$	$OCH(CH_3)_2$	$- 55.7$	Benzene	45)
$OCH(CH_3)_2$	$OCOCH_3$	$- 53.0$	Benzene	88)
$OCH(CH_3)_2$	**) $R'' = OCH(CH_3)_2$	$- 46.8$	Benzene	88)
$N(C_2H_5)_2$	OC_2H_5	$- 45.0$	Benzene	88)
$N(CH_3)_2$	$N(CH_3)_2$	$- 42.5$	Benzene	88)
$N(C_2H_5)_2$	**) $R'' = N(C_2H_5)_2$	$- 40.0$	Benzene	88)
OCH_3	$N(C_6H_5)_2$	$- 42.0$	Benzene	88)
$N(C_6H_5)_2$	$N(C_6H_5)_2$	$- 29.5$	pyridine	88)
CH_3	C_6H_5	$+ 6.5$	pyridine	85)

*) 2.4.6-Triphenylphosphorin for comparing.

**) $= -O-P$

5. Mass Spectra

The parent peaks are usually easily identified. Nevertheless, several 1.1-carbo- and 1.1-hetero-λ^5-phosphorins readily split off thermally the 1.1-substituents, so that here a strong peak corresponding to 2.4.6-triphenyl-λ^3-phosphorin (324) can be observed. This effect is particularly pronounced in 1.1-dialkylthio-2.4.6-triphenyl-λ^5-phosphorins, but also in 1.1-dibenzyl- and 1.1-diarylamino derivatives, and to a lesser extent in 1.1-bis-dialkylamino-λ^5-phosphorins. The mass spectral properties thus reflect the utility of potential protecting groups at the P atom of λ^5-phosphorins out of which λ^3-phosphorins can be regenerated. Kanter [92] has established the following sequence which reflects the increasing ease of thermal alkylthio cleavage from 1.1-dialkylthio-2.4.6-triphenyl-λ^5-phosphorins to form 2.4.6-triphenyl-λ^3-phosphorins:

ethylenedithio $\approx$ diethylthio $\ll$ dimethylthio $\approx$ propylenedithio.

Indeed, thermolysis of 1.1-dimethylthio-2.4.6-triphenyl-λ^5-phosphorin at 180 °C, neat or in toluene, affords preparative amounts of 2.4.6-triphenyl-λ^3-phosphorin in yields of 87% and 66%, respectively.

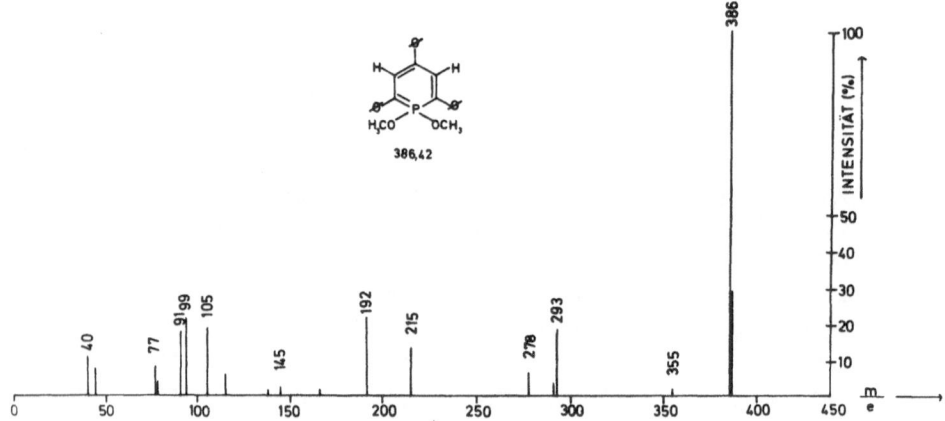

Fig. 33. Mass spectrum of 1.1-dimethoxy-2.4.6-triphenyl-λ^5-phosphorin

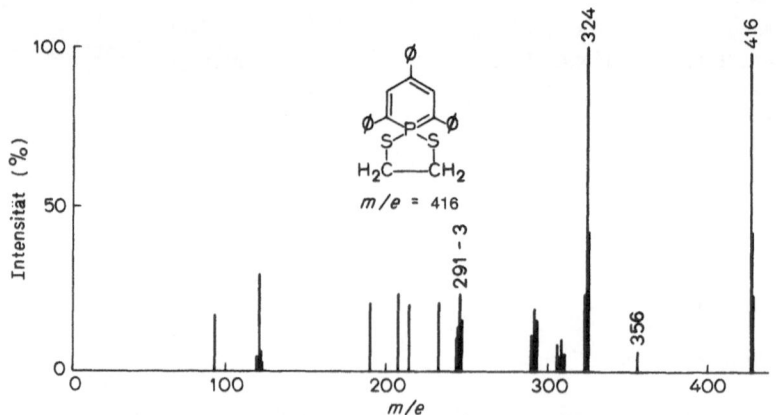

Fig. 34. Mass spectrum of 1.1-dithio-spiro-λ^5-phosphorin *170* (*n* = 2)

Apparently, smooth thermolysis requires a certain orientation of the alkylthio groups which allows easy disulfide formation.

The 2.4.6-triphenyl-λ^5-phosphorins often show the same mass spectral fragments as the 2.4.6-triphenyl-λ^3-phosphorins. For example, 1-phenyl-1-alkyl-2.4.6-triphenyl-λ^5-phosphorin has m/e peaks at 293 corresponding to the 1.2.4-triphenyl-cyclopentadiene-(1,3)-cation [86].

Figs. 33, 34 and 35 show the fragmentation patterns of several 1.1-hetero-2.4.6-triphenyl-λ^5-phosphorins. Note the large differences in the pattern of the two sulfur-λ^5-phosphorins in fig. 34 and 35 with respect to the mole- and the 2.4.6-triphenyl-λ^3-phosphorin-peak (m/e = 324).

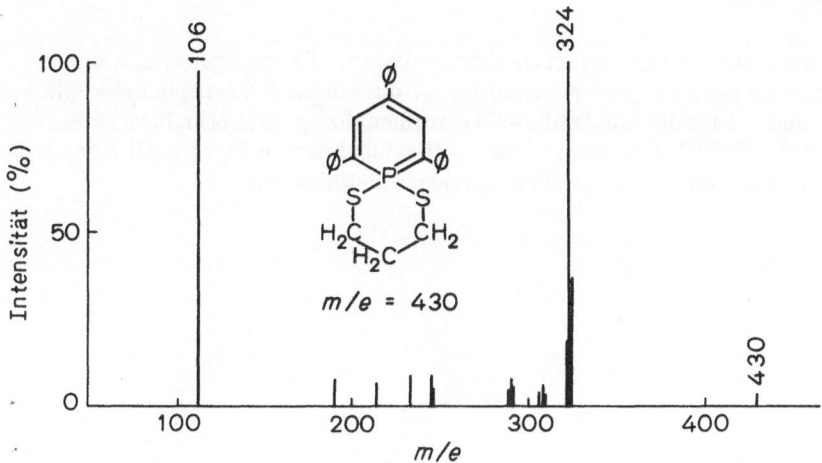

Fig. 35. Mass spectrum of 1.1-dithio-spiro-λ^5-phosphorin *170* (n = 3)

6. Dipole Moments

The dipole moment of 1.1-dimethoxy-2.4.6-triphenyl-λ^5-phosphorin *183* according to the Guggenheim method [103] was found to have a value of $\mu = 1.3$ D.*) A value of 2.7 D was found for 1.1-dimethoxy-2.4.6-diphenyl-4-(4'-chlorophenyl)-λ^5-phosphorin *184* [45].

The direction of the dipole moment is therefore established. Moreover, the data points to the partial "ylid" character of both λ^5-phosphorins, which is clearly much smaller in the delocalized "aromatic" π system than in open-chain ylids which have dipole moments of 5–7 D [104]. For comparison, the dipole moments of 2.4.6-triphenyl-λ^3-phosphorin *22* and 2.4.6-tri-tert-butyl-λ^3-phosphorin *24* were also determined.

---

*) A value of 1.2 D had previously been determined by Prof. Dr. H. Nöth, Marburg, now University of München.

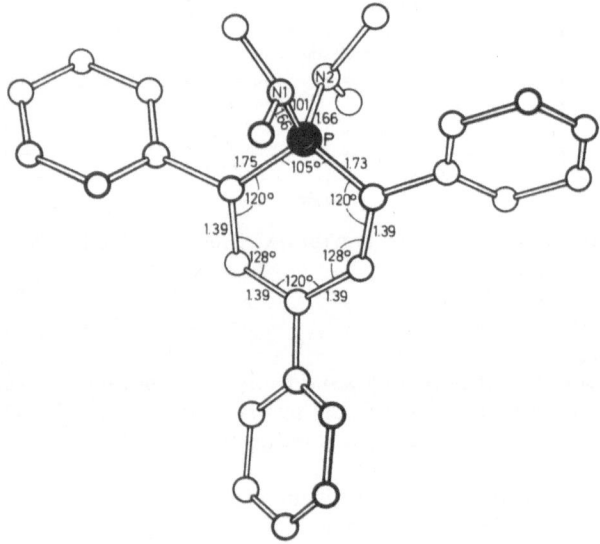

183	*184*	*22*	*24*
1.3 D	2.7 D	1.54 D	1.5 D

7. X-Ray Analysis

Three independent X-ray structure determinations of λ^5-phosphorins, i. e. 1.1-dimethyl-2.4.6-triphenyl-λ^5-phosphorin, 1.1-dimethoxy-2.4.6-triphenyl-λ^5-phosphorin and 1.1-bis-dimethylamino-2.4.6-triphenyl-λ^5-phosphorin, have been made [105, 106, 107]. The result of one of these is shown in Fig. 36. All X-ray structure determinations are in excellent agreement with another.

Fig. 36. Crystal structure of 1.1-bis-(dimethylamino)-2.4.6-tri-phenyl-λ^5-phosphorin [107]

In all three cases the λ^5-phosphorin ring is nearly planar, so that good π-orbital overlap is possible. Whereas the O–P–O angle is 93°, the N–P–N angle has a value of 102°. The C–P–C angles and the P–C bond lengths of λ^5-phosphorins are quite similar to those of λ^3-phosphorins (Figs. 12 and 13, pp. 35 and 36).

114

8. Photoelectron Spectrum

The PE spectrum of 1.1-dimethoxy-2.4.6-tri-tert-butyl-λ^5-phosphorin has been recorded by Schweig and Schäfer [108]; this is the only λ^5-phosphorin which has been studied so far (Fig. 37).

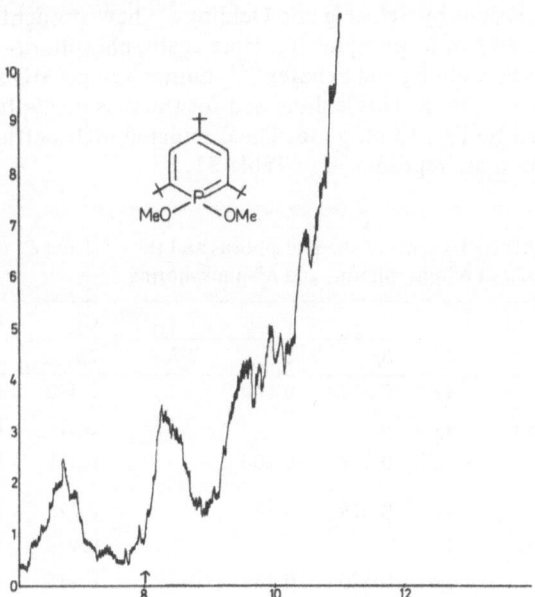

Fig 37. Photoelectronspectrum of 1.1-dimethoxy-2.4.6-tri-tert-butyl-λ^5-phosphorin

For comparision, one should inspect Fig. 14 with the PE spectrum of 2.4.6-tri-tert-butyl-λ^3-phosphorin *24* (p. 37). In going from *24* to 1.1-dimethoxy-2.4.6-tri-tert-butyl-λ^5-phosphorin the first band is shifted by 1.3 eV to lower ionization potential, while the second band remains at the same ionization potential. Due to the experimental intensity ratio of band 1: band 2 = 1:2 in *24*, the second band was attributed to the π_2 and n MOs. In 1.1-dimethoxy-2.4.6-tri-tert-butyl-phosphorin the second band does not include the n MO and has thus the same intensity as the first band. These observations experimentally support the orbital configuration of λ^3-phosphorins and λ^5-phosphorins predicted by Schweig and coworkers [53, 54].

D. Bonding in the λ^5-Phosphorin System

In analogy to calculations by Dewar and Whitehead [109] and Craig and Paddock [110] on phosphazenes, Märkl [82b] proposed a bonding model for λ^5-phosphorin ("non-classical phosphabenzene") with a linear combination of the d_{xy} and d_{yz} orbitals of phosphorus form $d_\pi{}^a$ and $d_\pi{}^b$ hybrid orbitals. These form d_π-p_π bonds with the neighboring sp^2-hybridized carbon atoms. In this model the P atom interrupts conjugation. In contrast, calculations by S. F. Mason [111] and R. Vilceanu [112] point to a fully aromatic system. They propose an extension of Hückel's Rule to include cyclic conjugated molecules having d_π-p_π bonds.

CNDO/2 calculations by Schweig and Oehling [54] have brought new insight concerning the bonding of λ^5-phosphorins. Here again, phosphorus does not block conjugation. Schweig, Oehling and Schäfer [53] found a unique MO sequence for this new heterocyclic system. This is discussed for the case of λ^3-phosphorin on p. 37 and illustrated by Fig. 15 on p. 38. The calculated MO coefficients of both bonding systems are reproduced in Table 31.

Table 31. π-AO coefficients of the phosphorus and the C–2 and C–6 atoms in the π-HOMO of λ^3-phosphorins and λ^5-phosphorins

		P			C_2	C_6
		$3pz$	$3d_{xz}$	$3d_{yz}$	$2pz$	$2pz$
λ^3-Phosphorin	4_1	0.217	0.140	–	0.342	0.342
	4_2	–	–	0.273	0.514	0.514
	4_3	0.599	0.200	–	0.379	0.379
λ^5-Phosphorin	4_1	0.019	0.168	–	0.275	0.275
	4_2	–	–	0.032	0.409	0.409
	4_3	0.320	0.312	–	0.445	0.445

These results show that the $3p_zAO$ of phosphorus contributes considerably to ring conjugation in λ^5-phosphorins [108]. The determining factor is that the highest occupied molecular orbital is of π type in both phosphorin systems. In λ^3-phosphorin the next lower MO is localized at the P atom to the extent of 60% (as an n MO). In the λ^5-phosphorin system this is not possible, which is in accordance with the observed PE spectral intensities [108] of Fig. 37, p. 115. The very different electron distribution of both λ^3- and λ^5-phosphorins in comparison to that of pyridine is in full accord with the chemistry of these classes of compounds:

E. Chemical Properties

1. Basicity; Addition of Alkyl or Acyl Ions

In contrast to λ^3-phosphorins, λ^5-phosphorins can be protonated. The basicity is very much influenced by the nature of the substituents R^1 and $R^{1\prime}$ at the phosphorus. 1.1-Dialkyl or 1.1-diaryl-λ^5-phosphorins are even protonated by aqueous HCl; the salts are deprotonated by aqueous NaOH. Strong acids in organic solvents, e. g. trifluoroacetic acid in hexane or benzene, (see p. 106), are required to protonate 1.1-dialkoxy-λ^5-phosphorins. Addition of tert-butoxide deprotonates the salt. By studying the NMR spectra of 1.1-dimethoxy-2.4.6-tris-pentadeuterophenyl-λ^5-phosphorin 185 in benzene solutions containing H and D-trifluoroacetic acid Städe [45)] could show that two different protonation products are formed in a ratio of 3:1. One product is the result of C—2 protonation 186 the other of C—4 protonation 187 (Fig. 38). Similar results were observed in the case of 1.1-bis-dimethylamino-2.4.6-triphenyl-λ^5-phosphorin [88)].

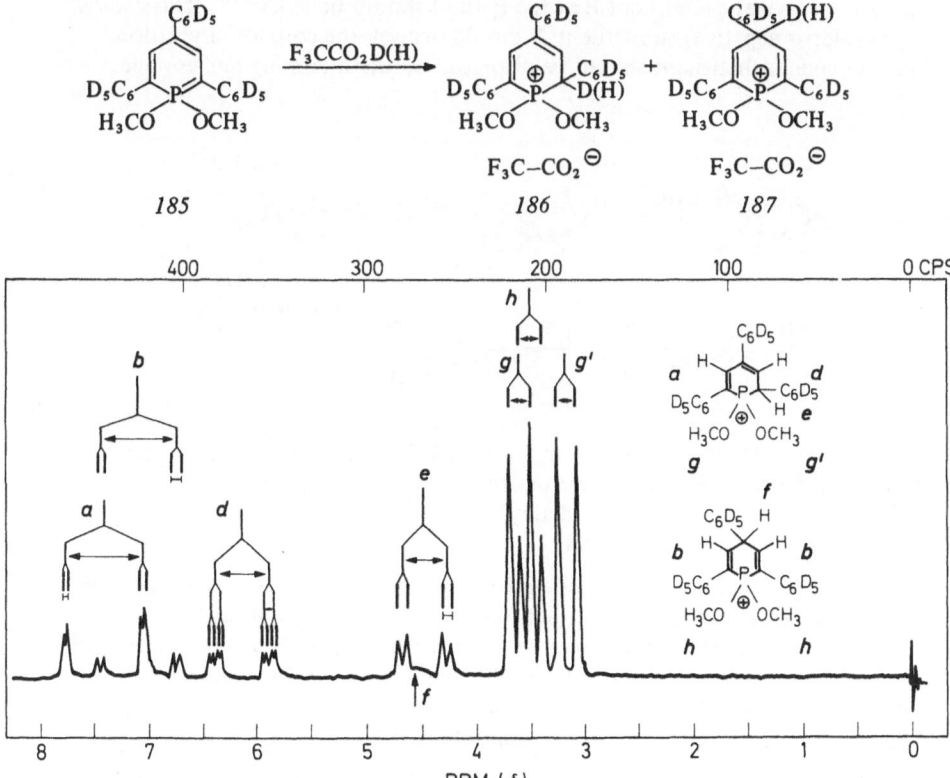

Fig. 38. 1H-NMR spectrum of 1.1-dimethoxy-2.4.6-tris-pentadeutero-λ^5-phosphorin in CF_3CO_2H.

117

Only those 1.1-diphenyl-λ^5-phosphorins in which the phosphorin ring is unsubstituted could be alkylated or acylated at the ring [90] (see p. 77). The 2.4.6-triphenylated 1.1-disubstituted λ^5-phosphorins cannot be alkylated by oxonium salts or acylated by acylchlorides under normal conditions.

2. Exchange of Substituents at Phosphorus by Nucleophilic Displacement

According to method K (p. 87), one of the amino groups in 1.1-bis-diarylamino- or 1.1-bis-dialkylamino-2.4.6-triphenyl-λ^5-phosphorins can be replaced by an alkoxy group by treatment with alcohols in the presence of trifluoroacetic acid. If thioalcohols are used, both amino groups are exchanged by alkylthio groups. Compounds of type *153* or *154* (p. 85) are also accessible to these exchange reactions. At present nothing definite can be said about the mechanism of these reactions. Protonation occurs at C−2 and C−4 of the ring. This has the effect of widening the C−P−C angle of the phosphorin ring of about 105−108°. In the transition state the C−2 and C−6 atoms remain in the equatorial position. The more bulky (or more negative) substituent B at the P atom should be axial [113]. The *smaller* (or less eletro negative) substituent A should occupy the equatorial position. Only the axial substituent should be displaced by the incoming nucleophile.

transition state

It is somewhat surprising that in 1.1-bis-dialkylamino-2.4.6-triphenyl-λ^5-phosphorin *both* dialkylamino groups are replaced by SR groups in the presence of thiols, whereas the 1-alkoxy-1-dialkylamino-λ^5-phosphorin fails to react at all under the same reaction conditions.

1-Alkoxy-1-acetoxy-2.4.6-triphenyl-λ^5-phosphorin *153* has two electrophilic centers, one at the carbonyl moiety and the other at the phosphorus atom. By varying the nucleophile and the second substituent at phosphorus R^1, Hettche [88, 98] was able to induce nucleophilic attack selectively either at the carbonyl group or at the P atom. In the former case, the anion of the 2-hydro-phosphorin acid ester *188b* is formed, in the latter a new λ^5-phosphorin *189*.

Attack at the carbonyl group with formation of the intense green-yellow fluorescent anion *188a* R = $CH(CH_3)_2$ occurs, for example, if sodium methanolate in benzene/methanol is used. In contrast, methyl magnesium bromide in ether attacks the phosphorus atom with ultimate formation of 1-methyl-1-iso-propoxy-2.4.6-tri-phenyl-λ^5-phosphorin 189 (R$'$ = CH$_3$, R = CH(CH$_3$)$_2$) in 60% yield. In the acetoxy-phosphorin series 153 in which the alkoxy groups R are sterically smaller substituents, such as OCH$_3$ or OC$_2$H$_5$, the yields of compounds 189 drop, while those of 188 rise. For example, addition of methyl magnesium bromide to 1-methoxy-1-acetoxy-2.4.6-triphenyl-λ^5-phosphorin 186 affords only 5% 1-methyl-1-methoxy-2.4.6-triphenyl-λ^5-phosphorin 189 (R = R$'$ = CH$_3$), the chief product (65%) being the anion $188a$ (R = CH$_3$). The more bulky tert-butoxy group seems to disturb attack at the carbonyl group but not so much at the phosphorus.

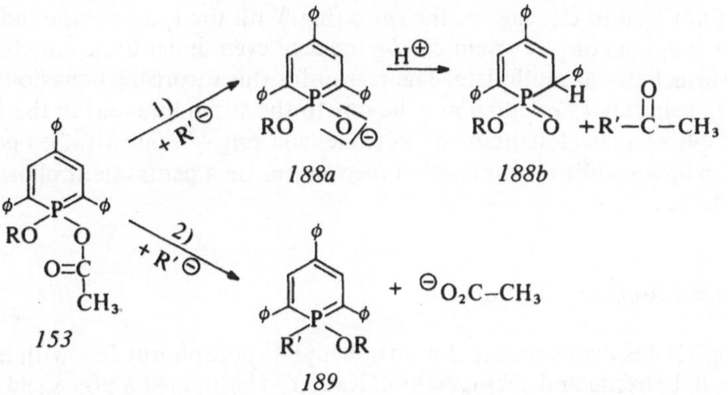

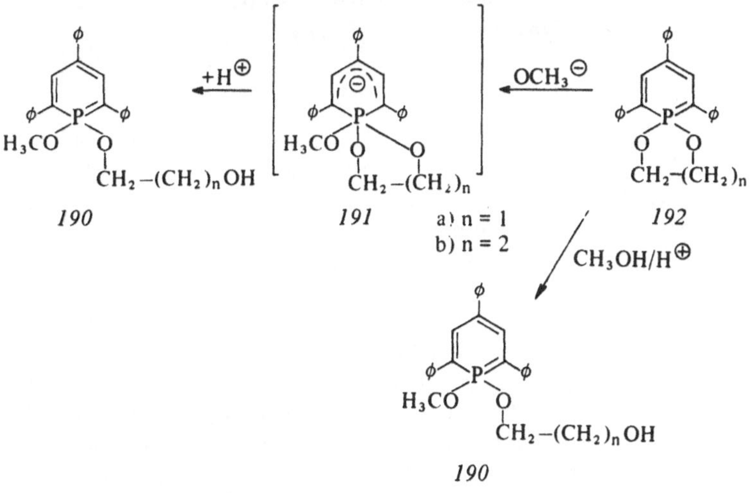

Whereas 1.1-dialkoxy-2.4.6-triphenyl-λ^5-phosphorins do not show any nucleophilic substitution reactions, one of the alkoxy groups of the spiro compounds $192a$ and b can easily be substituted with alcohols by an alkoxy group in the presence of trifluoroacetic acid to 190. Treatment with dimethylamine in acidic me-

dium does not lead to cleavage of the spiro ring. With the spiro compounds *192a* and *b* a nucleophilic displacement can be induced even under basic conditions, i. e. by such strong bases as methylate. The reason for this surprising behaviour of the spiro-λ^5-phosphorins *192a* and *b* may be due to the strain inherent in the 1.3-dioxa-2-λ^5-phospha-cyclopentane or -cyclohexane ring [45]. Substitution possibly occurs by an addition-elimination mechanism via a pentavalent phosphorus compound *191*.

3. Reactions with Boron Trifluoride

By treating 1.1-bis-dialkylamino-2.4.6-triphenyl-λ^5-phosphorin *164* with boron trifluoride in benzene and methanethiol, Kanter [92] obtained a 90% yield of 1-fluoro-1-dimethylamino-2.4.6-triphenyl-λ^5-phosphorin *172* (see p. 88).

4. Oxidation to Radical Cations

According to Städe [45] electrolytic oxidation of 1.1-dimethoxy-2.4.6-triphenyl-λ^5-phosphorin *193* with tetra-n-propylammonium perchlorate serving as the electrolyte leads to the formation of a λ^5-phosphorin radical cation *194* which is stable for several hours. Other 1.1-hetero-λ^5-phosphorins can be oxidized in the same manner to radical cations [63, 69].

The ESR spectra of these radical cations, just like those of the λ^3-phosphorin radical cations (see p. 42), show doublets with large phosphorus coupling constants a_p and well-resolved hyperfine structures due to hydrogen coupling. The a_p values as well as the general shape of the ESR spectra characterize the new radical cations. The hydrogen atoms of the methoxy groups have no significant influence on the spectra; 1.1-bis-trideuteromethoxy-2.4.6-triphenyl-λ^5-phosphorin yields the same ESR spectrum as the non-deuterated compound. Any coupling with these H atoms which are situated above and below the plane of the ring must be so small that it cannot be experimentally detected. Fig. 39 will serve as an example of an ESR spectra of one of these 1.1-hetero-2.4.6-triphenyl-λ^5-phosphorin radical cations, i.e. that of 1.1-bis-dimethylamino-2.4.6-triphenyl-λ^5-phosphorin radical cation. The ^{31}P coupling constants are summarized in Table 32.

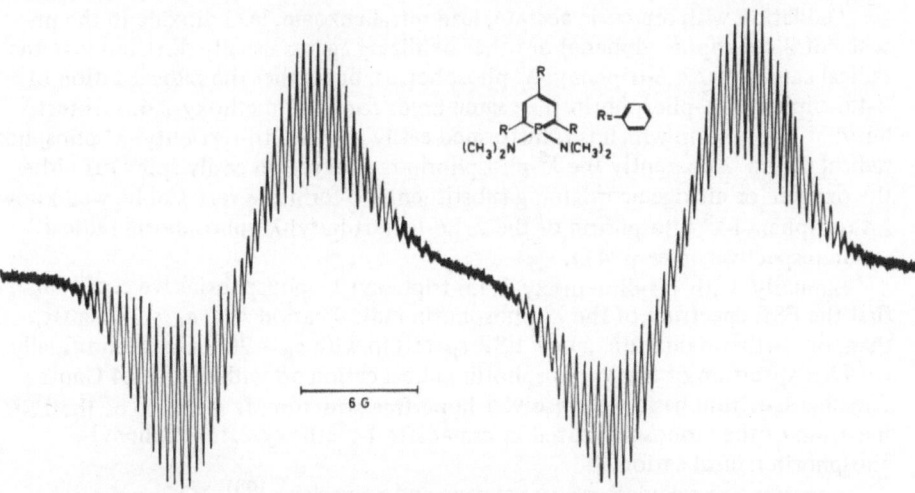

Fig. 39. ESR spectrum of 1.1-bis-dimethylamino-2.4.6-triphenyl-λ^5-phosphorin radical cation

Table 32. Coupling constants, total expansion and *g* values of
1.1-hetero-2.4.6-triphenyl-λ^5-phosphorin radical cations

	a_p (Gauss)	Total expansion (Gauss)	*g* value
[structure: CH_3O OCH_3]	18.0	35.9	2.002158
[structure: CH_3O $N(CH_3)_2$]	20.9	45.5	2.002152
[structure: $(CH_3)_2N$ $N(CH_3)_2$]	22.3	50	2.002286
[structure: P phosphorin]	23.4	41	2.002182

121

Oxidation with mercuric acetate, lead tetrabenzoate, lead dioxide in the presence of 2.4.6-triphenylphenol or other oxidizing agents usually does not give the radical cation of 2.4.6-triphenyl-λ^5-phosphorins, but rather the radical cation of 2.4.6-triphenyl-λ^3-phosphorin. The same holds for 1.1-dimethoxy-2.4.6-tri-tert-butyl-λ^5-phosphorin which is transformed easily to 2.4.6-tri-tert-butyl-λ^3-phosphorin radical cation. Apparently the λ^5-phosphorin-radical cation easily splits off either the oxygen- or nitrogen-containing substituents to form the very stable, well-known 2.4.6-triphenyl-λ^3-phosphorin or the 2.4.6-tri-tert-butyl-λ^3-phosphorin-radical cation respectively (see p. 41).

Similarly, with 1.1-dimethoxy-2.4.6-triphenyl-λ^5-phosphorins we could observe first the ESR spectrum of the λ^5-phosphorin radical cation *194*, $a_p = 18$ Gauss, then, on further oxidation, a new ESR spectrum with $a_p = 20.8$ Gauss and finally the ESR spectrum of the λ^3-phosphorin radical cation *58* with $a_p = 23.4$ Gauss. The new spectrum has a well resolved hyperfine structure. It seems to be the ESR spectrum of the monosubstituted intermediate 1-methoxy-2.4.6-triphenyl-phosphorin radical cation.

According to calculations by Schweig and coworkers [108], λ^5-phosphorins should be oxidized to the radical cation at a lower oxidation potential than the λ^3-phosphorins. Careful experiments on 1.1-dimethoxy-2.4.6-triphenyl-λ^5-phosphorin by Weber [63] confirmed these predictions. Accordingly, the 6π system of λ^5-phosphorin loses an electron more readily than that of λ^3-phosphorin; it is thus not the lone-electron pair at phosphorus, but rather the 6π-electron system which is responsible for the easy oxidation of the phosphorins to radical cations.

5. Cleavage of the 1.1-Substituents to Form λ^5-Phosphorins

Not only oxidation but also thermolysis of λ^5-phosphorins can lead to cleavage of both 1.1-substituents. According to Märkl [32] 1.1-dibenzyl-2.4.6-triphenyl-λ^5-phosphorin splits off 1.2-diphenylethane at temperatures higher than 220 °C to form 2.4.6-triphenyl-λ^3-phosphorins *22* (see p. 24). Other 1.1-carbo-2.4.6-triphenyl-λ^5-phosphorins also split off C groups at high temperatures to form 2.4.6-triphenyl-λ^3-phosphorin. The mechanism of cleavage may involve radicals.

Somewhat lower temperatures — about 180 °C — are required in the thermolysis of 1.1-bis-diphenylamino-2.4.6-triphenyl-λ^5-phosphorin. 1.1-Bis-tolylamino-2.4.6-triphenyl-λ^5-phosphorin *195* even reacts at 150 °C [88]. According to Hettche, cleavage is favored if the nitrogen radicals are trapped by such H-donating solvents

a) R = CH_3
b) R = C_2H_5
c) R = CH_2–CH_2 = R
d) R = CH_2–CH_2–CH_2 = R

as toluene or 1.3.5-triisopropyl benzene; besides 2.4.6-triphenyl-λ^3-phosphorin, diphenyl- and di-tolylamine are formed. This radical reaction can be viewed as the reverse of the previously described radical addition reaction (method E, p. 82).

Even more facile is the cleavage of thio groups from 1.1-dialkylthio- or diaryl-thio-2.4.6-triphenyl-λ^5-phosphorins *196* (see p. 88).

1.1-Bis-methylthio-2.4.6-triphenyl-λ^5-phosphorin *196a* is decomposed completely by refluxing in xylene for 48 hours. 2.4.6-triphenyl-λ^3-phosphorin *22* can be isolated on a preparative scale in 70% yield. However, it is easier merely to heat without solvent under nitrogen at 180 °C. In this case dimethyldisulfide is liberated, and the triphenyl-λ^3-phosphorin can be isolated in 87% yield. 1.1-Bis-ethylthio-2.4.6-triphenyl-λ^5-phosphorin *196b* requires somewhat higher temperatures. As the mass spectrum of the spiro-dithio-λ^5-phosphorin *196d* (Fig. 35, p. 112) shows, the thio groups are particularly easily cleaved. This may be due to the relief of ring strain. Preparative thermolysis leads to the same results.

The temperature of decomposition depends also on steric factors: The spiro compound *196 c* with the five-membered dithiaphospha ring is much more stable than the spiro compound *196 d* with the six-membered hetero ring. Only in the latter compound can a stable 1.2-dithia-cycloalkane be formed by radical cleavage Once again the preparative behaviour is in accord with the results of mass spectroscopy (p. 112).

In the nucleophilic displacement reaction of 1.1-bis-diethylamino-2.4.6-tri-phenyl-λ^5-phosphorin with thiophenol, the 1.1-diphenylthio-2.4.6-triphenyl-λ^5-phosphorin *197* does not survive the reaction temperature; only trace quantities are formed. Thiophenol in refluxing benzene is thus an excellent means of removing dialkylamino groups. Besides diphenyldisulfide and diethylamine, this procedure affords good yields of 2.4.6-triphenyl-λ^3-phosphorin *22*. 1-Phenoxy-1-di-ethylamino-2.4.6-triphenyl-λ^3-phosphorin *176* (see p. 89) is formed as a side product[92].

1.1-Dialkoxy- or 1.1-diaryloxy-2.4.6-triphenyl-λ^5-phosphorins are thermally very stable.

6. Cleavage of Alkyl Groups from 1.1-Dialkoxy-λ^5-phosphorins by BBr$_3$: 2-Hydro-phosphinic Acids

Treatment of 1.1-dimethoxy-2.4.6-triphenyl-λ^5-phosphorin *183* with boron tribo-mide in methylene chloride at 0 °C results in cleavage of the methyl group from the methoxy substituent at phosphorus. Hydrolysis affords 1-hydroxy-1-methoxy-2.4.6-tri-phenyl-λ^5-phosphorin *198* which exists almost completely in the tautomeric form, 2-hydro-phosphinic acid methyl ester *199*, as was shown on p. 60.

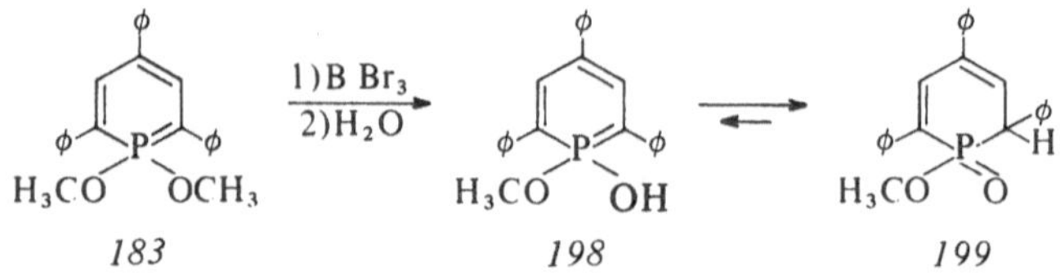

The BBr$_3$ reaction with 1.1-dimethoxy-2.4.6-di-tert-butyl-4-(4'-methoxyphenyl)-λ^5-phosphorin *200* leads to cleavage of both methoxy groups; in addition to the methoxy group at the phosphorus, the 4'-methoxy group is attacked. The 2-hydro-4-(4'-hydroxyphenyl)-phosphinic acid methyl ester *201* can be methylated with methyl iodide in methanol/sodium methylate at the phenolic group, leading to *202*, which can also be prepared by hydrogen peroxide oxidation of 2.6-di-tert-butyl-4-(4'-methoxy-phenyl)-λ^3-phosphorin *204* to *203*, followed by diazomethane methylation [44] (see Table 13, p. 61).

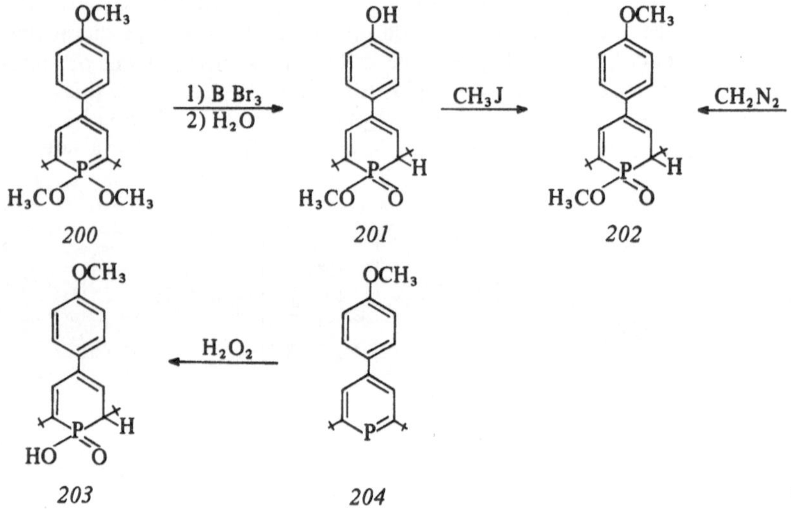

7. Oxidative Dealkylation of 1.1-Dialkoxy-λ^5-phosphorins: 4-Hydroxy-phosphinic Acids

Städe[45] discovered a smooth reaction by treating *183* with LiBr/H$_2$O$_2$ in 1.2-dimethoxyethane/glacial acetic acid (25:1). The reaction occurs at 20–25 °C and is completed within several minutes. Besides the alkyl bromide, 4-hydroxy-phosphinic acid ester *206* is formed. We suppose that a hydroxyl cation adds to the 4-position of the 1.1-dialkoxy-λ^5-phosphorin, forming the phosphonium ion *205*. This intermediate is attacked by the bromide ion at the alkoxy group, thereby yielding the alkyl bromide and the 4-hydroxy-phosphinic acid ester *206*. This reaction can be applied to all 1.1-dialkoxy-λ^5-phosphorins (see also Table 11, p. 57).

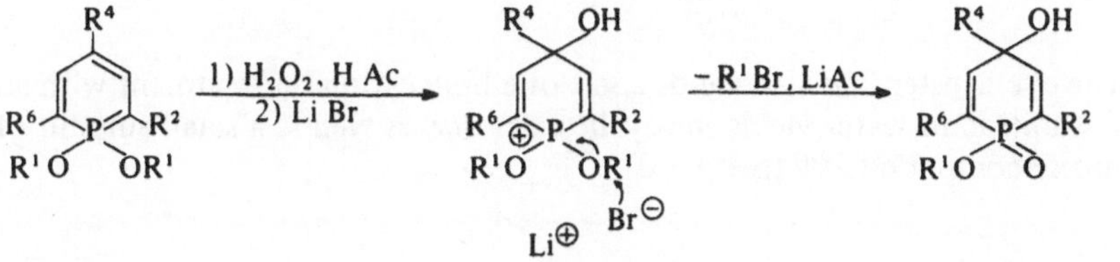

R^1, = alkyl
R^2, R^4, R^6 = aryl

Similar treatment of 1.1-spiro compounds leads to different results depending upon the number of atoms in the spiro ring. The five-membered compound *207 a* yields the cyclic 2-oxyphosphinic acid ester *208 a* (60%) and only 8% of the 4-hydroxyphosphinic acid ester *209a* (R = H). Acetic acid ist not necessary for these transformations. *207b* and *c* (R = D, CH$_3$) react in a similar manner:

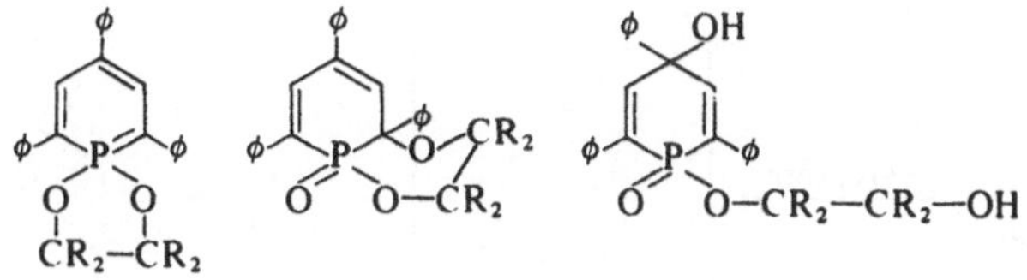

R = H, D, CH$_3$

The six-membered spiro compound *210* reacts with H$_2$O$_2$/LiBr/HAc to form besides some byproducts the stereoisomeric 4-hydroxy-3'-bromopropyl-phosphinic esters *211 Z* and *E*. *) This experiment of Städe [45] strengthens the proposed mechanism for oxidative dealkylation formulated above.

Alternative ways to prepare 4-hydroxy-phosphinic acids of type *206* have been described previously (Table 11, p. 57).

---

*) Nomenclature see p. 49.

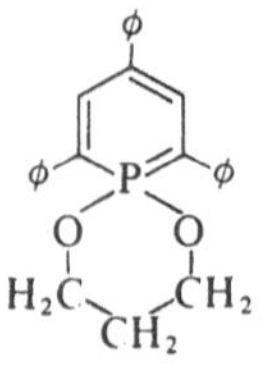

210 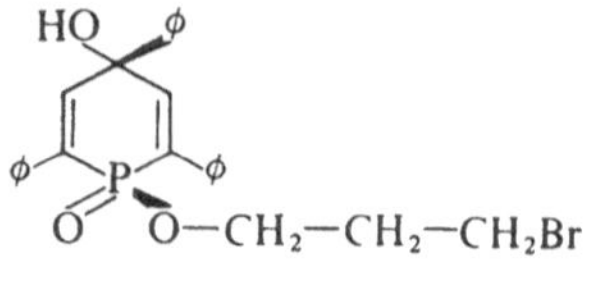   211 Z 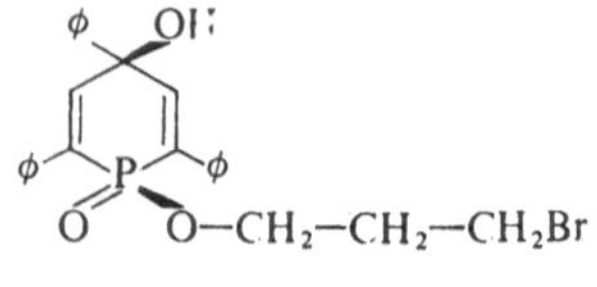   211 E

The cyclic ester *208* also yields a salt of a blue cation on treatment with strong acids. Addition of water yields the cyclic ester *208* as well as a small amount of the 4-hydroxy compound *209* (see p. 50).

8. Photoreactions

a) Photosensitized Oxidation

Completely different results from those obtained in the photooxidation of 2.4.6-tri-tert-butyl-λ^3-phosphorin *24* (p. 54) are obtained in the photooxidation of 1.1-dimethoxy-2.4.6-tri-tert-butyl-λ^5-phosphorin *183*, as Schaffer [75] has found. In this case the 2-hydroxy-endoxy-phosphinic acid methyl ester *213* can be isolated in about 20% yield. Its formation can be explained by assuming "normal" 1.4-addition to *212* as the primary product which is transformed to *213* by hydrolytic ring cleavage, followed by loss of methanol.

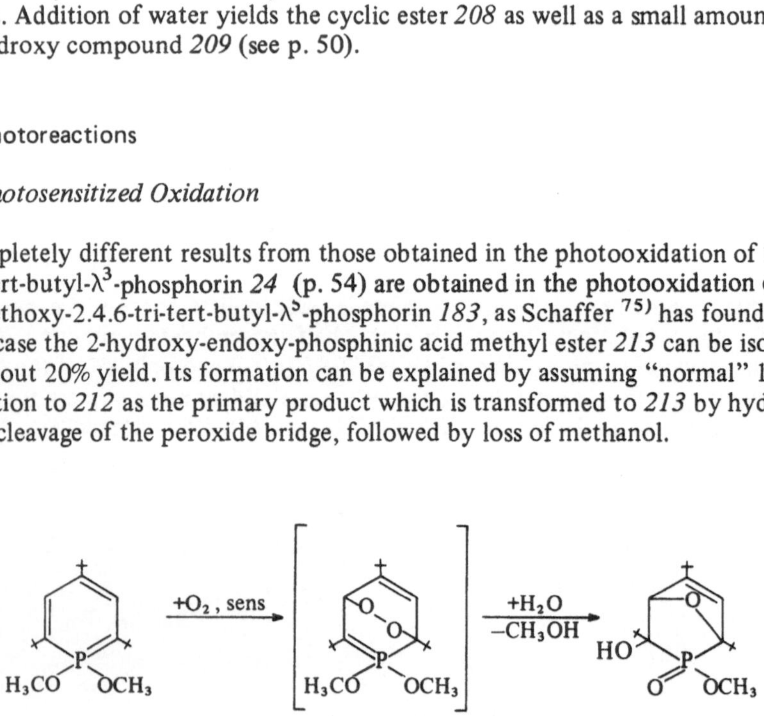

183   212   213

b) Photoisomerization of Cyclic Phosphinic Esters

2.4.6-Trisubstituted λ^3-phosphorins or λ^5-phosphorins can be isolated unchanged even after long periods of irradiation when oxygen is excluded. Städe [45] discovered that cyclic phosphinic acid esters *208 a–c* which contain a cyclic butadiene (1.3)-moiety, photochemically rearrange smoothly to the tricyclic compounds *214 a–c*. All analytical and spectroscopic data are in full agreement with the tricyclic structure.

126

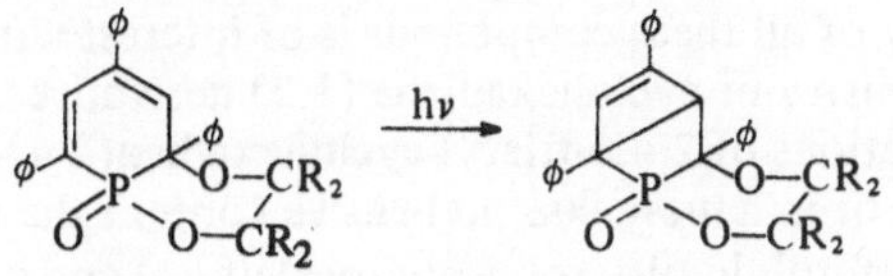

$$208 \qquad\qquad 214$$

a) R : H
b) R : D
c) R : CH$_3$

Treatment of 214 a with trifluoroacetic acid and water causes take-up of one mole H$_2$O. The new compounds have not yet been identified.

Hettche[88] investigated irradiation of compounds of type 215a and b. He isolated the crystalline photoproducts 216a and b which are isomers of 215a and b.

The exact structure is still under discussion and seems to be analogous to the structure of 214.

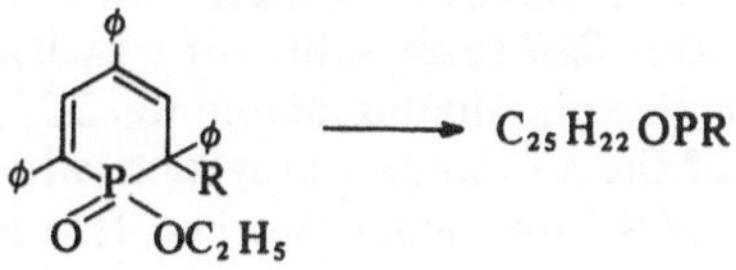

$$215 \qquad\qquad 216$$

a : R = H
b : R = C$_2$H$_5$

Compound 217 when irradiated with a high-pressure mercury lamp takes another course, as Constenla [101] has found. In the first step the acetyl group migrates to position 2, but the stereochemistry of this photoproduct 218 has not yet been determined. This step is analogous to a photochemical Fries rearrangement. In the next step the phosphorus residue is split off and 2.4.6-triphenyltoluene 219 can be isolated.

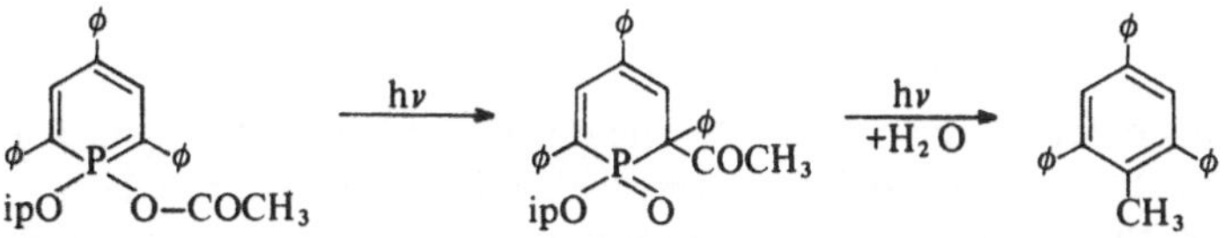

$$217 \qquad\qquad 218 \qquad\qquad 219$$

The photochemistry of all these compounds is of interest with respect to the well-known photochemistry of cyclohexadiene (1.3) derivatives, for instance, the photochemical reactions of 2.4.6-triaryl-cyclohexadien(2.4)-on-ol(2) series of Perst and Dimroth[114], or in a more comprehensive context the photochemical transformation of ergosterol, lumisterol, isopyrocalciferol and pyrocalciferol which were first studied by Windaus and Dimroth and fully investigated by Veluz, Havinga and Dauben [115].

9. Hydride Cleavage from 1.1-Hetero-4-methyl-2.6-diphenyl-λ⁵-phosphorins to Form Stable Carbenium Phosphonium Ions

a) Carbenium-phosphonium-oxonium Salts

4- or 2-Methyl-substituted λ⁵-phosphorins, in contrast to 4- or 2-methyl-pyridines or pyridinium cations show no tendency to split off a proton from the methyl group with bases or to react with carbonyl compounds in aldol-type condensations.

However, with triphenylcarbenium-tetrafluoroborate, 1.1-dimethoxy-2.6-diphenyl-4-methyl-λ⁵-phosphorin 220 easily splits off a hydride ion to form a resonance-stabilized carbenium-phosphonium-oxonium-ion 221, as Schäfer has found [67]. The high electron density of the λ⁵-phosphorin system, which is in exellent agreement with the theoretical model of Schweig and coworkers (see p. 115), facilitates this interesting reaction.

a+c: 7.3–7.8 ppm (m mit d, 12H)
 b: 2.18 ppm (s, 3H)
 d: 3.24 ppm (d, $J = 11.5$ Hz, 6H)

a′: 8.18 ppm (d, $J = 42$ Hz, 2H)
b′: 6.84 ppm (s, 2H)
c′: 7.8 ppm (s, 10H)
d′: 3.90 ppm (d, $J = 13$ Hz, 6H)

In going from 220 to the cation 221, the position of the NMR signals of the protons are shifted, as expected, to lower field. This effect is particularly pronounced in the case of the methyl protons (b) in 220 as compared to the methylene protons in 221; see Figs. 40 and 41.

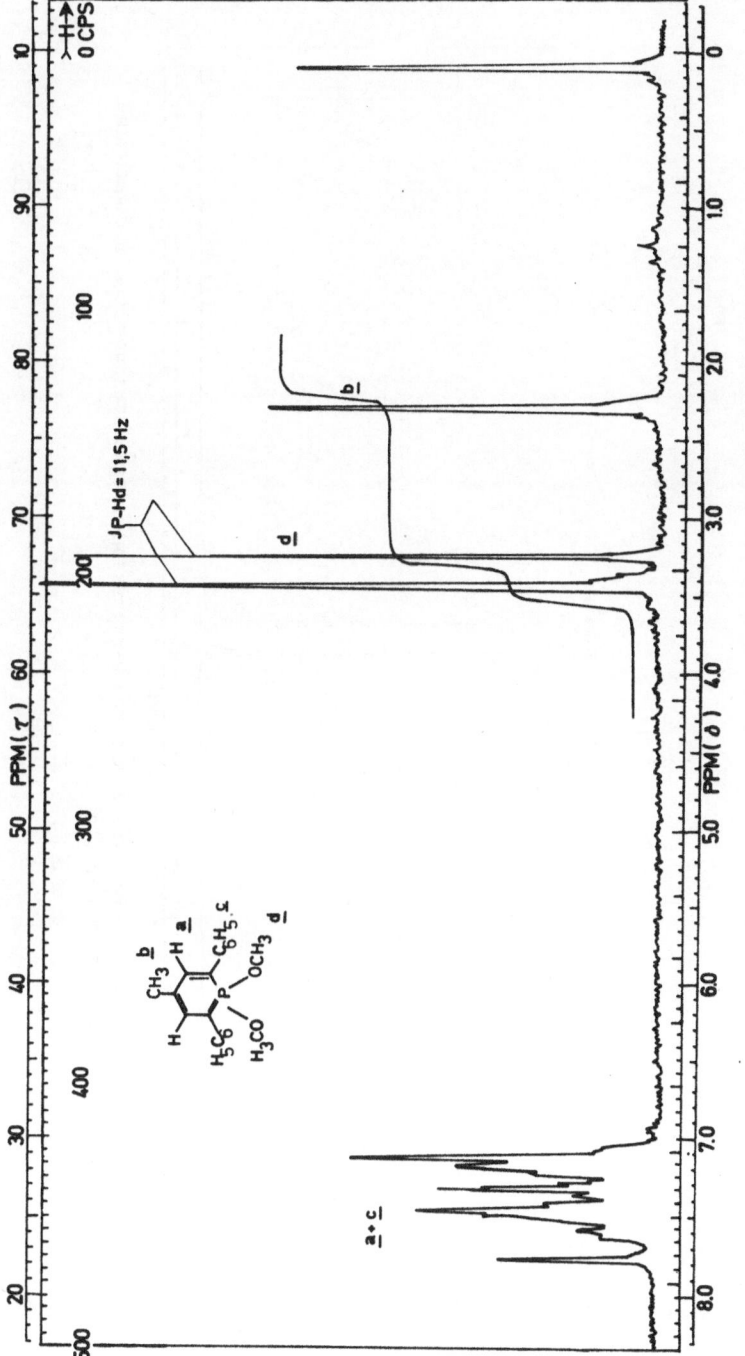

Fig. 40. ^{1}H-NMR spectrum of 1.1-dimethoxy-4-methyl-2.6-diphenyl-λ^5-phosphorin in CDCl$_3$

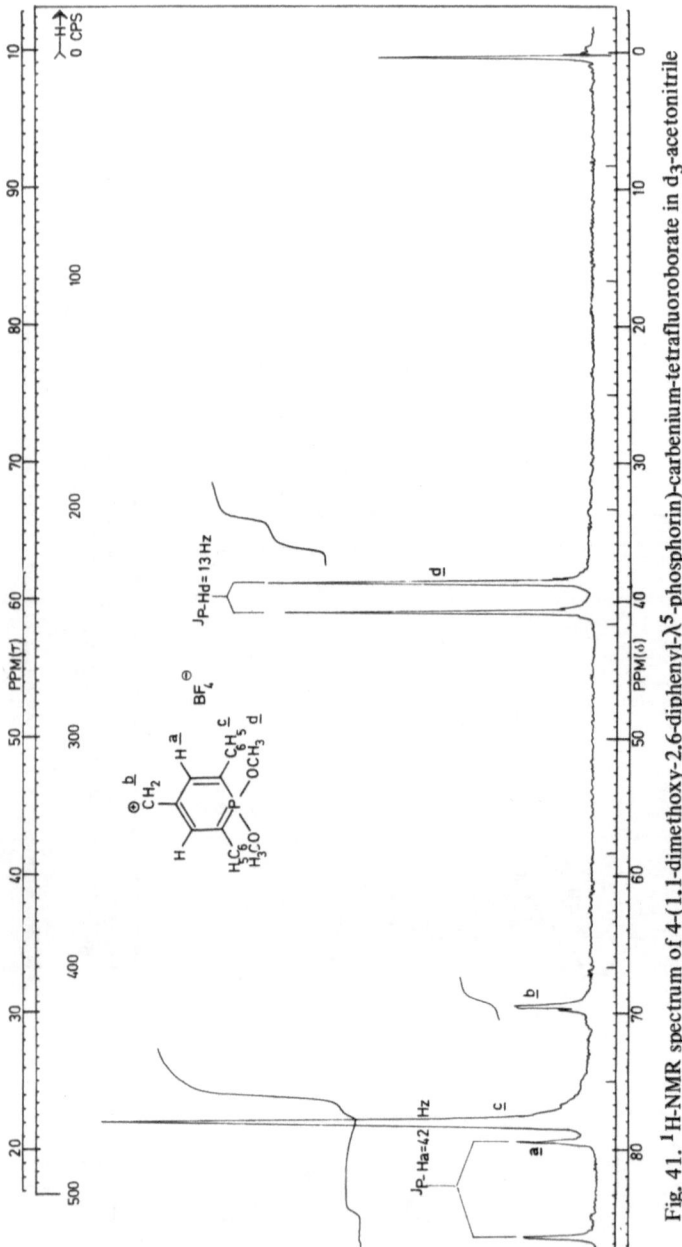

Fig. 41. ¹H-NMR spectrum of 4-(1,1-dimethoxy-2,6-diphenyl-λ⁵-phosphorin)-carbenium-tetrafluoroborate in d₃-acetonitrile triphenylcarbeniumtetrafluoroborate. (Table

In a similar manner *222* is transformed into the crystalline tetrafluoroborate *223*.

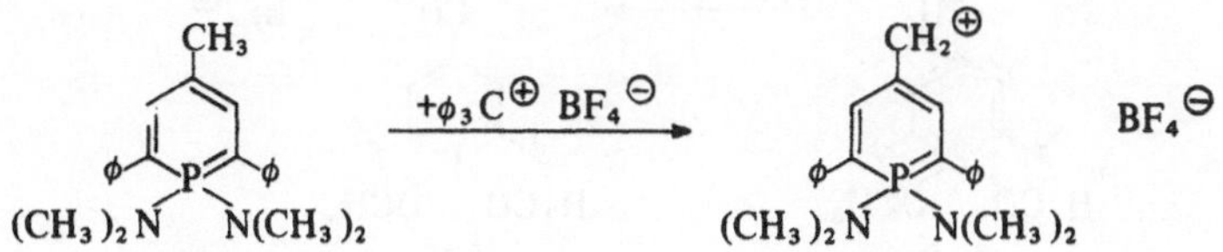

| | *222* | | | *223* | |

Schäfer and Pohl [67, 96, 97)] succeeded in transforming a number of 1.1-dialkoxy-λ^5-phosphorins having different CHR_2 groups at position C−4 into crystalline, analytically pure 4-substituted-methylene-λ^5-phosphonium-tetrafluoroborates of type *221* by the action of triphenylcarbenium-tetrafluoroborate (Table 33).

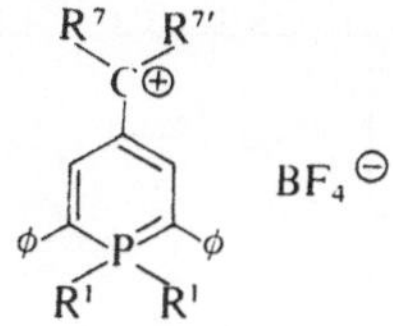

Table 33. 4-(1.1-Hetero-λ^5-phosphorin)-carbenium-tetrafluoroborates

R^1	R^7	$R^{7'}$	m. p. °C	Lit.
OCH_3	H	H	160	67)
OC_2H_5	H	H	120	67)
OC_6H_5	H	H	−	123)
OCH_3	H	CH_3	134−5	67)
OCH_3	H	C_6H_5	148−9	123)
OCH_3	CH_3	CH_3	157−8	123)
$N(CH_3)_2$	H	H	135 (decomp.)	67)

Another and still more stabilized carbenium-phosphonium-oxonium-tetrafluoroborate *225* can be prepared when bis-4.4-'(1.1-dimethoxy-2.6-diphenyl-λ^5-phosphorin)-methane *224* is treated with triphenylcarbenium-tetrafluoroborate.

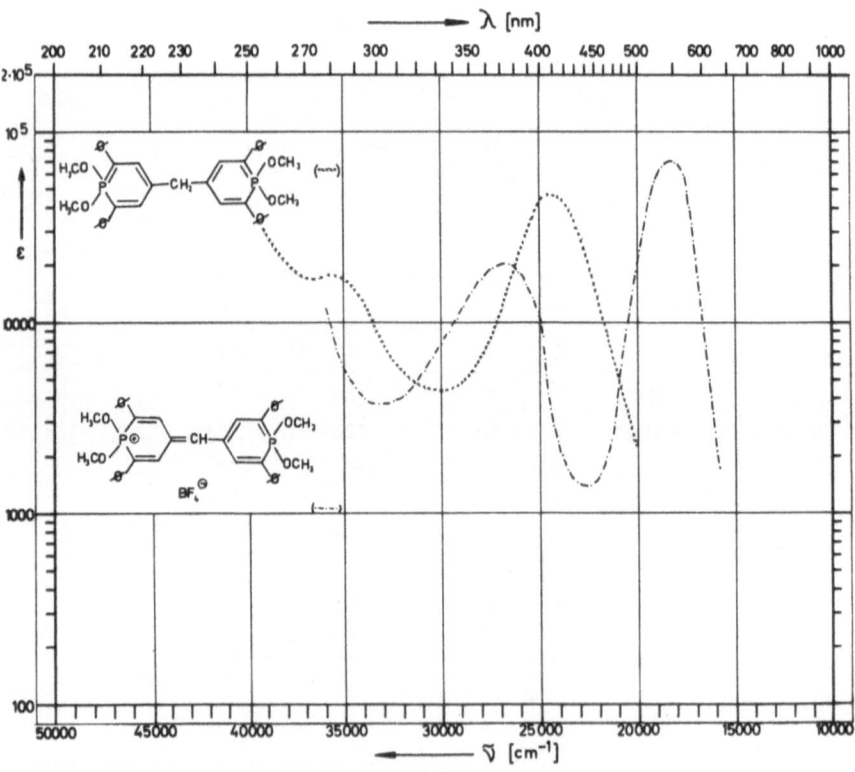

The deep red salt *225* belongs to the group of cyanine dyes containing phosphorus as cationic end groups. The long-wave absorption maximum with the very high extinction coefficient of ϵ = 68000 lies at λ = 545 nm, the second maximum at 375 nm (ϵ = 10300). In contrast *224* absorbs at 412 nm (ϵ = 34100) and at 285 nm (ϵ = 12000),(Fig. 42).

Fig. 42. UV spectra of *224* in cyclohexane and of *225* in CH_2Cl_2

Cyanine dyes of a similar type to *225* were also obtained by Märkl [32)] via a completely different route (see p. 77).

b) Reactions of the Carbenium-phosphonium-oxonium Salts with Nucleophiles

The resonance forms of the cation *221* and *225* suggest that nucleophiles can attack at three different positions:

1. At the carbenium atom C–4′ (formation of a new 4′-substituted λ^5-phosphorin or 4′-substituted *224*).
2. At the P atom (formation of a pentacoordinated phosphorus compound).
3. At the alkyl group of the O-alkyl substituent (cleavage of an alkyl cation and formation of an unsaturated phosphinic acid ester).

Of these possibilities, only reactions 1. and 3. have been observed.

Concerning 1:

A number of nucleophiles X smoothly add to the carbenium ion center of *221* (in acetonitrile) [67)]:

a) Hydride ion to form *226 a* which is identical with compound *220* m. p. 141–142 °C.

b) Cyano ion, to form *226 b* m. p. 102–103 °C.

c) Thiocyanate ion, to form a mixture (which we could not separate on a preparative scale) of the thiocyanate *226 c* and the isothiocyanate *226 c* m. p. 140 °C.

d) 4-Dimethylamino-benzene, to form the coupling product *226 d*, m. p. 113 °C [122)].

$221 \xrightarrow{+X^{\ominus}}$

226

a) $X^{\ominus}$ = H (= *220*)
b)   = CN
c)   = CNS or SCN

d) X = ⟨benzene ring⟩–N(CH₃)₂

$225 \xrightarrow{+X^{\ominus}}$

227

a) $X^{\ominus}$ = H (= 224)
b)   = CN
c)   = OCH₃
d)   = OC₂H₅

In this way λ^5-phosphorins with functional groups in the side chain can be prepared.

Similar additions of nucleophiles to *225* lead to the compounds *227 a–d*. In this case OCH_3^- or $OC_2H_5^-$ could also be added (*227 c* and *d*). With acids these compounds regenerate the cation of *225* [67, 96, 97].

Concerning 3:

In both cations *221* and *225* the alkyl groups are easily split off by halide ions. Whereas addition to the carbenium ion leads in a *kinetically* controlled[116] reaction to the *addition* products *226* and *227*, this type of reaction leads to the *thermodynamic* stable unsaturated *phosphinic acid esters 228* and *229*.

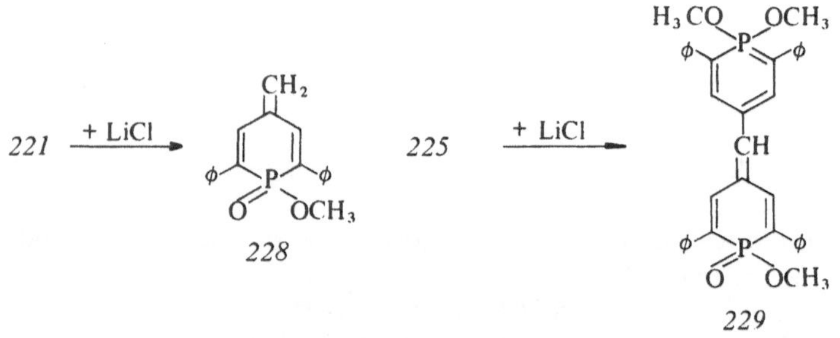

The reaction proceeds very easily with $Cl^{\ominus}$, $Br^{\ominus}$ and $J^{\ominus}$-ions. In the case of *221*, alcoholate leads to compound *228*, whereas under cautious condition with *225* the addition products *227c* and *d* can be isolated.

The ^{1}H-NMR spectra of the λ^5-phosphorins *226* and *227*, are very different from the ^{1}H-NMR spectra of *228* and *229*. Whereas in the λ^5-phosphorins *226* and *227* the signals of the two ring protons at C–3 and C–5 are found at low field (for instance *226 b* at $\delta = 7.38$ ppm, $J_{P-H} = 33$ Hz), in *228* they are shifted to higher field: $\delta = 6.71$ ppm $J_{P-H} = 33.5$ Hz, (fig. 43). Also the UV spectra are shifted to short waves: *228*: $\lambda_{max} = 328$ nm, $\epsilon = 3290$; *229* $\lambda_{max} = 479$ nm, $\epsilon = 31900$.

When the addition products *226 c* are refluxed in acetonitrile solution, they split off methylthiocyanate (which was not isolated) and are transformed into the thermodynamically stable 4-methylene-phosphinic methyl ester *228* [67].

c) Electrophilic Substitution at C–4

A most interesting reaction was found when the tetrafluoroborate *221* in acetonitrile was treated with water or with weak aqueous bases. Instead of the expected hydroxymethyl derivative *230*, only the 4.4'-bis-(1.1-dimethoxy-2.6-diphenyl-

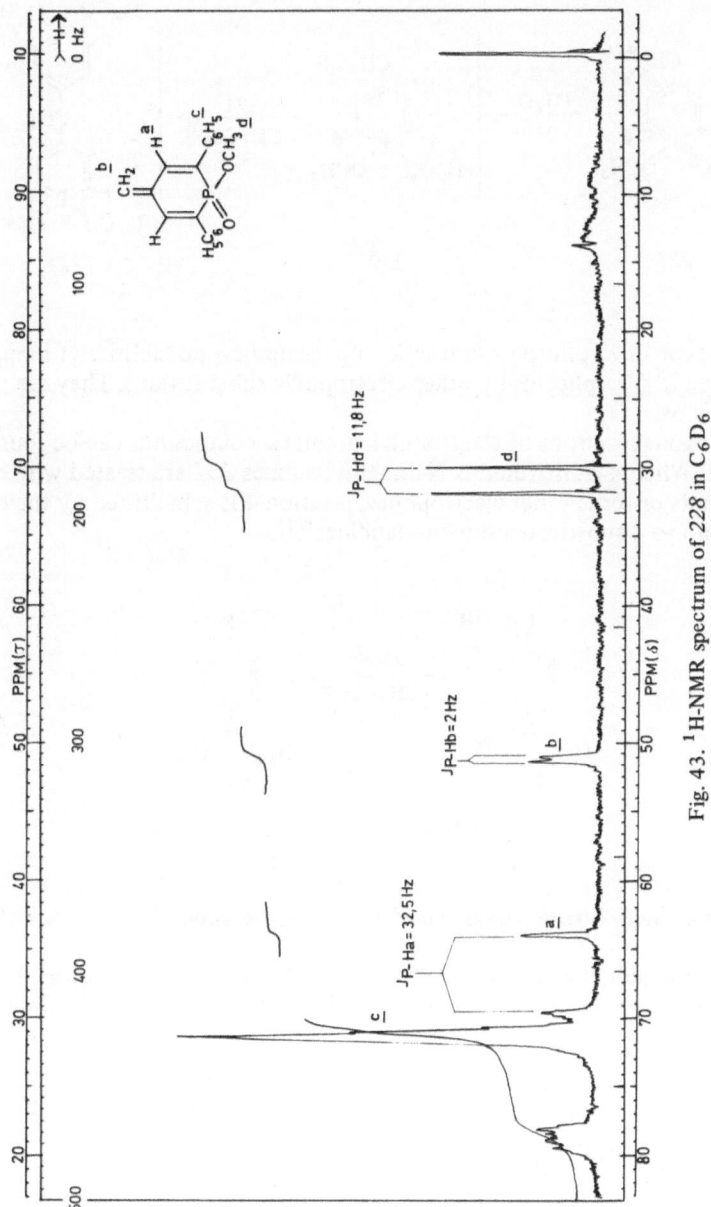

Fig. 43. ^{1}H-NMR spectrum of 228 in C_6D_6

λ^5-phosphorin)-methane 224 m. p. 141–142 °C, in 86% yield, could be isolated. This reaction proceeds by an electrophilic attack of the cation 221 on the 4-hydroxymethyl-λ^5-phosphorin intermediate 230, CH_2OH^+ being split off. Indeed, formaldehyde can be isolated [67, 97].

135

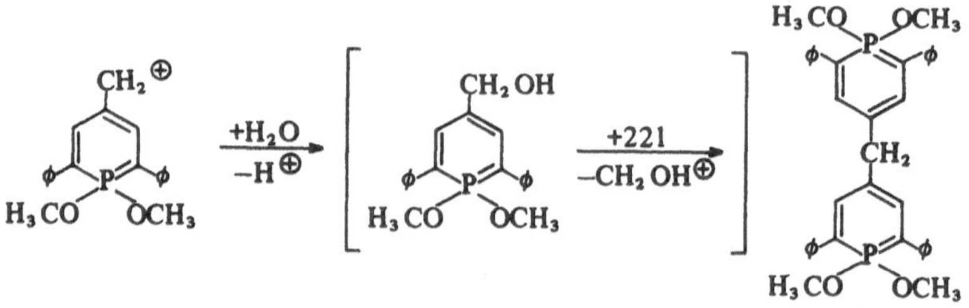

221   230   224

This type of λ^5-phosphorin reaction opens up new possiblities of preparing 4-substituted λ^5-phosphorins by other electrophilic substitutions. They are now being studied by us.

Analogous reactions of electron-rich aromatic compounds can be found in the literature: When 4-substituted-N.N-dimethylanilines *231* are treated with diazonium salts or some other electrophiles, position 4 is substituted by the electrophile group to 4-substituted dimethylanilines[117].

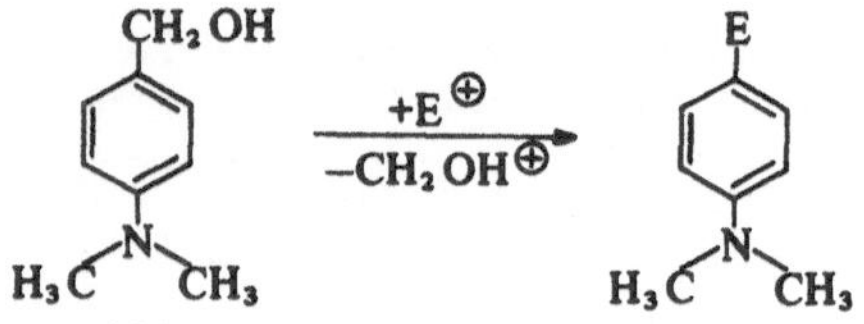

231   232

$$E = N_2 \, aryl^\oplus, NO_2^\oplus \, u. \, a.$$

d) Reactions with Strong Bases – a λ^5-Phosphorin-carbene?

It seemed possible that treatment of cation *233a* with a strong proton abstracting base could led to a resonance stabilized carben[118,119] *234a* (R = H). Pohl[43] by

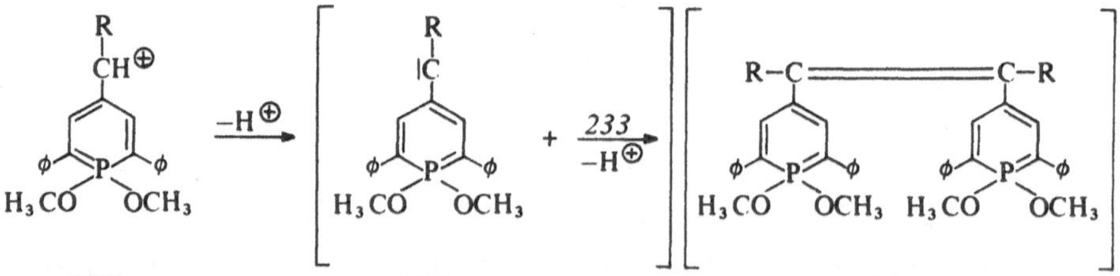

233   234   235

a) R = H
b) R = CH₃

136

reacting *233a* with diisopropylethylamine was able to isolate a crystalline substance, which was formed by three phosphorin unities. From the analytical and spectroscopic data we suppose an aliphatic or a cycloaliphatic structure *236*. Under the same conditions *233b* forms an analogous compound, whereas 1.1-dimethoxy-2.6-diphenyl-4-dimethylcarbenium-λ^5-phosphorin-tetrafluoroborate (*233*, R and H = CH_3) with bases produces the monomeric olefin (1.1-dimethoxy-2.6-diphenyl-4-isopropenyl-λ^5-phosphorin).

We suppose that *233a* resp. *233b* with the base loses a proton to give the carben *234a* resp. *234b* which immediately reacts with the cation losing another proton to *235a* resp. *235b*. This electron rich olefin which could not be isolated, reacts again with the cation *233a* resp. *233b*, losing a CH_3-cation for final product *236a* resp. *236b*.

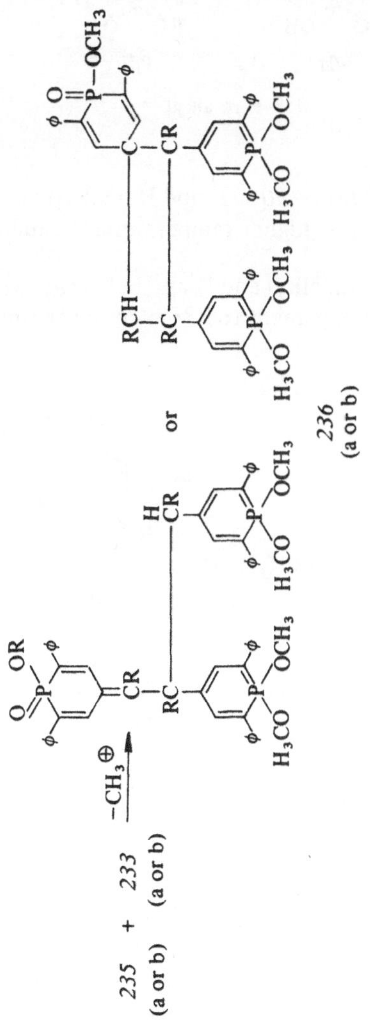

10. Rearrangements

a) [1.7] Acyl Rearrangement

As mentioned on p. 60, 1.1-dihydroxy-λ^5-phosphorins *90a* (R=H) exist almost entirely in the 2-hydro-phosphinic acid form *90b*. The same holds for the 1-alkoxy-1-hydroxy-λ^5-phosphorins (R = alkyl).

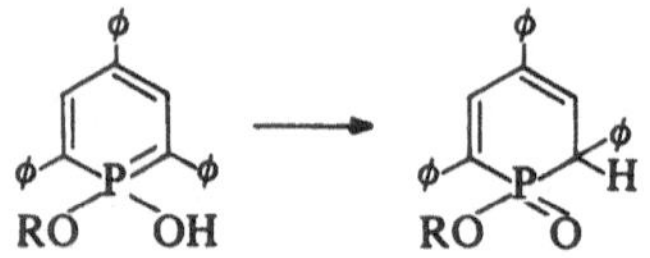

90a   *90b*

R = H or alkyl

1.1-Dialkoxy-λ^5-phosphorins (*90 a*, R and H = alkyl) at temperatures such as 200 °C do not rearrange, but at higher temperatures a similar rearrangement cannot be excluded.

However, if in *90 a* R is an alkyl and H an acyl group, as in compound *237*, Hettche found that the rearrangement to *238* takes place smoothly at 60–70 °C [88, 98].

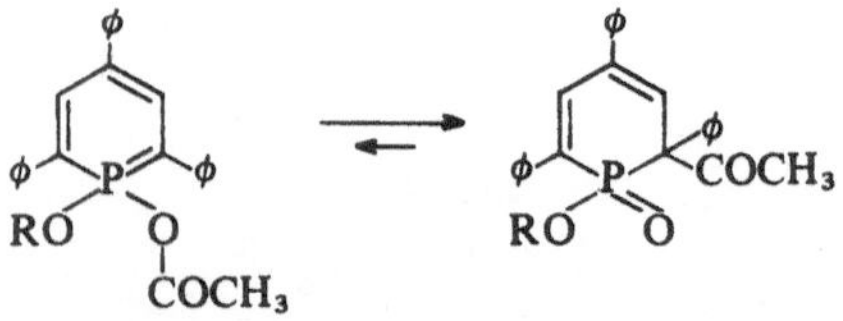

237   *238*

The reaction was investigated in detail by Constenla [101]. It leads to an equilibrium in favor of *238* which can be reached from both sides

$$K_{82,1°} = 15.7; K_{101,8°} = 13.9 \text{ (decaline)}$$

The reaction is exothermic with a small entropy gain from *237* to *238*:
$\Delta G^0 = -1.9$ kcal/mol; $\Delta H^0 = -1.6$ kcal/mol; $\Delta S^0 = 1$ eu.

Since the absorption spectra of *237* and *238* are rather different, the kinetics can be easily studied by UV spectroscopy. As Fig. 44 shows, the rearrangement is very clean, no side-products could be found.

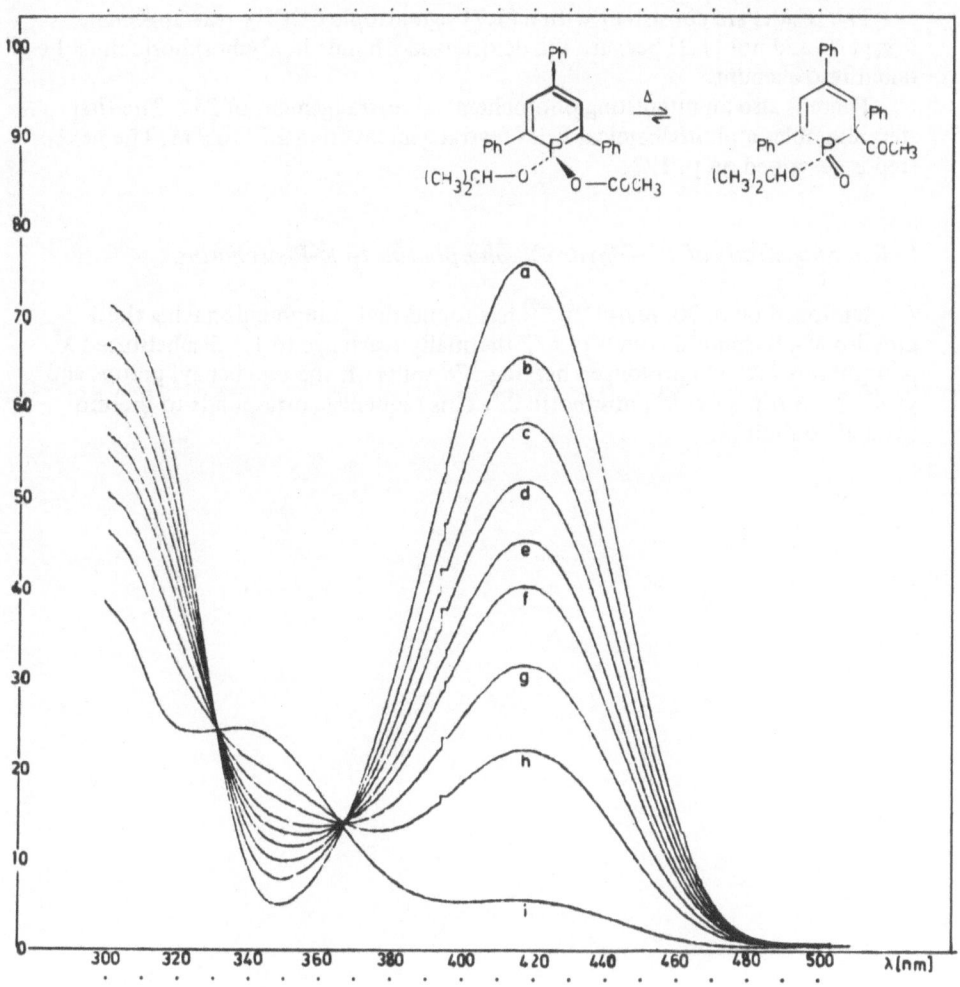

Fig. 44. UV spectra of *237* → *238* (R = CH(CH₃)₂) at 72.5 °C in dioxane 9 x 10⁻⁵ molar

a) at the beginning (3 min)
b) after 75 min
c) after 135 min   f) after 315 min
d) after 195 min   g) after 435 min
e) after 255 min   h) after 615 min
                          i) after 1400 min

First-order kinetics were obtained. At 91.5⁰ in decaline $\overrightarrow{k}$ = 26,1 x 10⁻⁵ sec⁻¹, $\overleftarrow{k}$ = 1.8 x 10⁻⁵ sec⁻¹ were found. Since the rate is nearly independent of solvent polarity (six solvents with E_T values between 31.2 and 46.0 were used), a cyclic transition state seems probable. The activation parameters were found to be $\Delta G^{\ddagger}$ = 27 kcal/mol, $\Delta H^{\ddagger}$ = 24.4 kcal/mol, $\Delta S^{\ddagger}$ = –8.6 e.u.

The results are consistent with a [1.7] sigmatropic $COCH_3$ rearrangement; it is [1.7] and not [1.3] because the delocalized π bonds in λ^5-phosphorin must be taken into account.

There is also an interesting photochemical rearrangement of *237*. The first step resembles a photochemical Fries rearrangement from *237* to *238*. The next step is described on p. 127.

b) Rearrangements of 1.2-Dihydro-λ^3-phosphorins to λ^5-Phosphorins

As mentioned on p. 90, Märkl [86, 49] has found that compounds having the 1.2-dihydro-λ^3-phosphorin structure *177* thermally rearrange to 1.1-disubstituted λ^5-phosphorins *178*. On prolonged heating *178* splits off the two benzyl groups and yields 2.4.6-triphenyl-λ^3-phosphorin *22*. This sequence corresponds to thermodynamic stabilities.

V. Outlook

The chemistry of phosphorus compounds with a delocalized P–C double bond proves to be very versatile. Whereas the physical properties of phosphamethin-cyanines are similar to the corresponding methin- or azamethin-cyanines, their chemical properties are distinguished by the higher reactivity of the phosphorus atom and the phosphorus-carbon double bond.

λ^3-phosphorins have physical properties which are rather similar to those of pyridines. But the chemistry of λ^3-phosphorins is very different, due mainly to the phosphorus atom which can easily lose one electron to produce a stable radical cation, or accept one or more electrons to yield a radical anion, dianion or radical trianion. Nucleophiles add to stable λ^4-phosphorin anions. In contrast to pyridine chemistry, no stable λ^4-phosphorinium compound (corresponding to a N-alkyl-pyridinium salt) could be isolated. Instead the electron shell of phosphorus is enlarged by addition of an electrophile yielding a λ^5-phosphorine derivative.

λ^5-phosphorins constitute a novel and very versatile class of heterocyclic compounds, the properties of which can be varied considerably by substituents at both the phosphorus and the ring C atoms. The ylid properties are fully suppressed by π-electron delocalization over the entire ring.

λ^3- and especially λ^5-phosphorins are electron-rich aromatic compounds, comparable with aniline, whereas pyridine and pyridinium ions are electron-poor and are comparable to nitrobenzene. Many chemical properties can be easily understood once this fact is taken into account.

The high stability of λ^3-phosphorins and λ^5-phosphorins suggests that additional "aromatic" ring systems can be prepared, having more than one phosphorus atom or heteroatoms other than phosphorus.

The physiological and pharmacological properties of the new compounds described in this review have not yet been investigated. Thus, nothing can be said concerning potential applications.

Acknowledgement

Our work was generously supported by the Deutsche Forschungsgemeinschaft, the Fonds der Chemischen Industrie of West Germany and the Deutsche Akademische Austauschdienst (M. Constenla).

It is a pleasure to acknowledge the many experimental and theoretical contributions of my coworkers mentioned in this review, particularly those of Dr. A. Hettche, who also critically read this paper. I also would like to tank Dr. M. Reetz for translation into English and for critical comments.

VI. Literature

1) Hudson, R. F.: Structure and mechanism in organo-phosphorus chemistry. London and New York: Academic Press 1965.
2) a) Dimroth, K., Hoffmann, P.: Angew. Chem. 76, 433 (1964); Angew. Chem. Intern. Ed. 3, 384 (1964).
   b) Dimroth, K., Hoffmann, P.: Chem. Ber. 99, 1325 (1966).
3) a) Allmann, R. A.: Angew. Chem. 77, 134 (1965); Angew. Chem. Intern. Ed. 4, 150 (1965);
   b) Allmann, R. A.: Chem. Ber. 99, 1332 (1966).
4) Patterson, A. M., Capell, L. T., Walker, D. F.: The ring index. J. Am. Chem. Soc. 281, 37, 1960.
5) Märkl, G.: Angew. Chem. 78, 907 (1966); Angew. Chem. Intern. Ed. 5, 846 (1966).
6) a) Märkl, G.: Angew. Chem. 75, 669 (1963); Angew. Chem. Intern. Ed. 2, 620 (1963);
   b) Märkl, G.: Angew. Chem. 77, 1109 (1965); Angew. Chem. Intern. Ed. 4, 1023 (1965).
7) a) Wittig, G., Geissler, G.: Liebigs Ann. Chem. 580, 44 (1953); Wittig, G., Maerker, A.: Chem. Ber. 97, 747 (1964);
   b) Johnson, A. W.: Ylid chemistry. New York – London: Academic Press 1966.
8) Cragg, R. H.: Essays in Chemistry. Vol. 1, p. 77 London and New-York: Academic Press 1970. Shaw, R. A., Fitzsimmons, A. B. W., Smith, B. C.: Chem. Rev. 62, 247 (1962); Fluck, E.: Topics in phosphorus chemistry. Vol. 4, p. 291. New York: Interscience 1967.
9) a) Greif, N.: Diss. Univ. Marburg 1967.
   b) Greif, N., Dimroth, K.: Unpublished results.
10) Märkl, G., Lieb, F.: Tetrahedron Letters 1967, 3489.
11) a) Klapproth, A.: Dipl. Arb. Univ. Marburg 1970 and Diss. Univ. Marburg 1972;
    b) Klapproth, A., Dimroth, K.: Unpublished results.
12) Kuhn, H.: Helv. Chim. Acta 31, 1441 (1948); 34, 1308 (1951); Z. Electrochem. 53, 165 (1949).
13) Hüning, S.: Optische Anregung organischer Systeme, S. 208. Weinheim: Verlag Chemie 1966.
14) Dimroth, K., Hoffmann, P.: French patent 1437938, Chem. Abstr. 66, P 11879g (1967).
15) Brooker, L. G. S., Sklar, A. L., Cressmann, H. W., Keyes, G. H., Smith, L. A., Sprague, R. H., Lare, E. van, Zandt, G. van, White, F. L., Williams, W. W.: J. Am. Chem. Soc. 67, 1875 (1945); Brooker, L. G. S., Sprague, R. H., Cressmann, H. W. J.: J. Am. Chem. Soc. 67, 1889 (1945).
16) Brunings, K. J., Corwin, A. H.: J. Am. Chem. Soc. 64, 593 (1942).
17) Kawada, I., Allmann, R.: Angew. Chem. 80, 40 (1968); Angew. Chem. Intern. Ed. 6, 69 (1968).
18) Wheatley, P. J.: J. Chem. Soc. London 1959, 3245, 4096.
19) Daly, J. J.: J. Chem. Soc. London 1964, 3799.
20) Treibs, A., Zimmer-Galler, R.: Liebigs Ann. Chem. 627, 166 (1959).
21) Dimroth, K., Bräuninger, G., Hoffmann, P.: Unpublished results, (1955, 1965).
22) Price, Ch. C., Parasaran, T., Lakshminarayan, T. V.: J. Am. Chem. Soc. 88, 1034 (1966).

Literature

23) Dimroth, K., Greif, N., Städe, W., Steuber, F. W.: Angew. Chem. *79*, 725 (1967); Angew. Chem. Intern. Ed. *6*,711 (1967).
24) Dimroth, K., Mach, W.: Angew. Chem. *80*, 489 (1968); Angew. Chem. Intern. Ed. *7*, 460 (1968).
25) Märkl. G., Lieb, F., Merz, A.: Angew. Chem. *79*, 475 (1967); Angew. Chem. Intern. Ed. *6*, 458 (1967).
26) Märkl, G., Lieb, F., Merz, A.: Angew. Chem. *79*, 947 (1967); Angew. Chem. Intern. Ed. *6*, 944 (1967).
27) Krafft, W.: Diss. Univ. Marburg 1962; Dimroth, K., Neubauer, G.: Angew. Chem. *69*, 720 (1957); Dimroth, K., Krafft, W., Wolf, K. H.: Nitro compounds, Proc. Intern. Symp. Warschau 1963, Pergamon Press *1964*, 361.
28) Dilthey, W., Fischer, J.: Ber. Dtsch. Chem. Ges. *56*, 1012 (1923); *57*, 1653 (1924); Schneider, W.: Liebigs. Ann. Chem. *432*, 297, especially page 317; Buck, J. S., Heilbron, J. M.: J. Chem. Soc. London *123*, 2521 (1923); Wizinger, R., Wagner, K.: Helv. Chim. Acta *34*, 2290 (1951); Balaban, A. T., Schroth, W., Fischer, G.: Advan. Heterocyclic Chem. *10*, 241 (1969).
29) Dimroth, K.: Angew. Chem. *72*, 331 (1960); Dimroth, K., Wolf, K.H.: Newer methods of preparative chemistry, Vol. 3, p. 357. New York: Academic Press 1964.
30) Steuber, F. W.: Lecture Chemiedozententagung Hamburg 1968.
31) Steuber, F. W., Dimroth, K.: Unpublished results.
32) Märkl, G.: Chimie Organique du Phosphore, Paris 1969; Edition du Centre National de la Recherche Sci. Paris (VIIe) 1970, No. 182, p. 295.
33) Koe, P. De, Bickelhaupt, F.: Angew. Chem. *79*, 533 (1967); Angew. Chem. Intern. Ed. *6*, 567 (1967).
34) Koe, P. de, Veen, R. van, Bickelhaupt, F.: Angew. Chem. *80*, 486 (1968); Angew. Chem. Intern. Ed. *7*, 465 (1968).
35) Koe, P. de, Bickelhaupt, F.: Angew. Chem. *80*, 912 (1968); Angew. Chem. Intern. Ed. *7*, 889 (1968).
36) Dimroth, K., Odenwälder, H.: Chem. Ber. *104*, 2984 (1971).
37) Chatzidakis, A.: Diss. Univ. Marburg 1969; Chatzidakis, A., Dimroth, K.: Unpublished results.
38) Ashe, A. J.: J. Am. Chem. Soc. *93*, 3293, 6690 (1971).
39) Ashe, A. J., Shu, P.: J. Am. Chem. Soc. *93*, 1804 (1971).
40) Jutzi, P., Deuchert, K.: Angew. Chem. *81*, 1051 (1969); Angew. Chem. Intern. Ed. *8*, 991 (1969); Maier, L., Seyferth, D., Stone, F. G. A., Rochow, E. G.: J. Am. Chem. Soc. *79*, 5884 (1957).
41) Oedinger, H., Kabbe, H. J., Möller, F., Eiter, K.: Chem. Ber. *99*, 2012 (1966); Oedinger, H., Möller, F.: Angew. Chem. *79*, 53 (1967); Angew. Chem. Intern. Ed. *6*, 76 (1967).
42) Bart, J. C., Daly, J. J.: Angew. Chem. *80*, 843 (1968); Angew. Chem. Intern. Ed. *7* 811 (1968).
43) Pohl, H. H.: Dipl.-Arbeit, Univ. Marburg (1971); Pohl, H. H., Dimroth, K.: Unpublished results.
44) Mach, W.: Diss. Univ. Marburg (1968); Mach, W., Dimroth, K.: Unpublished results.
45) Städe, W.: Diss. Univ. Marburg (1968); Städe, W., Dimroth, K.: Unpublished results.
46) Tolmachev, A. I., Kozlov, E. S.: Zh. Obshch. Khim. *37*, 1922 (1967); Chem. Abstr. *68*, 105298^h (1968); Zhungietu, G. I., Chukhrii, F. N., Tolmachev, A. I.: Zh, Obshch. Khim. *1970*, 590; Chem. Abstr. *74*, 22956 (1971).
47) Schoeler, U.: Dipl.-Arbeit, Univ. Marburg (1968).
48) Märkl, G., Fischer, D. E., Olbrich, H.: Tetrahedron Letters *1970*, 645
49) Märkl, G.: 20 Jahre Fonds der Chemischen Industrie p. 113, Frankfurt/M. Karlstr. 21, 1970.
50) Fischer, W., Hellner, E., Chatzidakis, A., Dimroth, K.: Tetrahedron Letters *1968*, 6227.
51) Daly, J. J.: J. Chem. Soc. *1964*, 3799.
52) Daly, J. J., Wheatley, P. J.: J. Chem. Soc. A, *1966*, 1703; Daly, J. J.: J. Chem. Soc. A, *1967*, 1913.

53) Oehling, H., Schäfer, W., Schweig, A.: Angew. Chem. *83*, 723 (1971); Angew. Chem. Intern. Ed. *11*, 656 (1971).

54) Oehling, H., Schweig, A.: Tetrahedron Letters *1970*, 4941.

55) Klages, F., Träger, H.: Chem. Ber. *86*, 1327 (1953); Degani, J., Fochi, R., Vincenzi, C.: Gazz. Chim. Ital. *94*, 203 (1964); Dimroth, K., Kinzebach, W., Soyka, M.: Chem. Ber. *99*, 2351 (1966).

55a) Oehling, H., Schweig, A.: Phosphorus *1*, 203 (1971).

56) Nöth, H., Deberitz, J.: Unpublished results.

57) Deberitz, J.: Dipl.-Arbeit, Univ. Marburg/Lahn 1969.

58) Deberitz, J., Nöth, H.: Chem. Ber. *103*, 2541 (1970).

59) Vahrenkamp. H., Nöth, H.: Chem. Ber. *105*, 1148 (1972).

60) Dimroth, K., Greif, N., Perst, H., Steuber, F. W.: Angew. Chem. *79*, 58 (1967); Angew. Chem. Intern. Ed. *6*, 85 (1967).

61) Dimroth, K., Städe, W.: Angew. Chem. *80*, 966 (1968); Angew. Chem. Intern. Ed. *7*, 881 (1968).

62) Isenberg, I., Baird, S. L.: J. Am. Chem. Soc. *84*, 3803 (1962).

63) Weber, H.: Dipl.-Arbeit,Univ. Marburg/Lahn 1971. Weber, H., Dimroth, K.: Unpublished results.

64) Dimroth, K., Berndt, A., Bär, F., Volland, R., Schweig, A.: Angew. Chem. *79*, 69 (1967); Angew. Chem. Intern. Ed. *6*, 34 (1967).

65) McConnell, H. M.: J. Chem. Phys. *24*, 764 (1956); Karplus, M. and Fraenkel, G. K.: ibid. *35*, 1312 (1961).

66) Thomson, C., Kilcast, D.: Chem. Commun. *1971*, 214.

67) Schäfer, W.: Diss. Univ. Marburg/Lahn, 1972; Schäfer, W., Dimroth, K.: Unpublished results.

68) Dimroth, K., Steuber, F. W.: Angew. Chem. *79*, 410 (1967); Angew. Chem. Intern. Ed. *6*, 445 (1967).

69) Dimroth, K.: Chimie Organique du Phosphore, Paris 1969. Edition du Centre National de la Recherche Sci. Paris (VIIe) *1970*, No. 182, p. 139

70) Dimroth, K., Vogel, K., Mach, W., Schoeler, U.: Angew. Chem. *80*, 359 (1968); Angew. Chem. Intern. Ed. *7*, 371 (1968).

71) Garbisch, E. W., Patterson, D. B.: J. Am. Chem. Soc. *85*, 3228 (1963).

72) Lemieux, R. U., Kullnig, R. K., Bernstein, H. J., Schneider, W. G.: J. Am. Chem. Soc. *80*, 6098 (1958).

73) Bieman, K.: Mass spectrometry, organic chemical applications, p. 145. New York: McGraw Hill 1962.

74) Doak, G. O., Freedman, L. D., Levy, J. B.: J. Org. Chem. *29*, 2382 (1964).

75) Dimroth, K., Chatzidakis, A., Schaffer, O.: Angew. Chem. *84*, 526 (1972); Angew. Chem. Intern. Ed. *11*, 506 (1972).

76) Skorianetz, W., Schulte-Elte, K. H., Ohloff, G.: Helv. Chim. Acta *54*, 1913 (1971); Schulte-Elte, K. H., Willhalm, B., Ohloff, G.: Angew. Chem. *81*, 1045 (1969); Angew. Chem. Intern. Ed. *8*, 985 (1969).

77) Märkl, G., Lieb, F., Merz, A.: Angew. Chem. *79*, 59 (1967); Angew. Chem. Intern. Ed. *6*, 86 (1967).

78) Märkl, G., Lieb, F.: Angew. Chem. 80, 702 (1968); Angew. Chem. Intern. Ed. 7, 733 (1968).

79) Lieb, F.: Diss. Univ. Würzburg 1969 (lit. [78]).

80) Märkl, G., Lieb, F.: Martin, C.: Tetrahedron Letters *1971*, 1249.

81) Märkl, G., Merz, A.: Tetrahedron Letters *1971*, 1269.

82) a) Märkl, G.: Angew. Chem. *75*, 168, 669 (1963); Angew. Chem. Intern. Ed. *2*, 153, 479 (1963).
   b) Review: Märkl, G.: Angew. Chem. *77*, 1109 (1965); Angew. Chem. Intern. Ed. *4*, 1023 (1965).

83) Johnson, A. W.: Ylid chemistry. New York and London: Academic Press 1966; Bestmann, H., Zimmermann, R.: Fortschr. chem. Forsch. *20*, 1 (1971).

84) Märkl, G.: Tetrahedron Letters *1961*, 807.

145

Literature

85) Märkl, G., Merz, A.: Tetrahedron Letters *1963*, 3611.
86) Märkl, G., Merz, A.: Tetrahedron Letters *1969*, 1231.
87) Märkl, G., Merz, A.: Tetrahedron Letters *1971*, 1215.
88) a) Hettche, A.: Diss. Univ. Marburg 1971
    b) Hettche, A., Dimroth, K.: Unpublished results; see also [94, 95, 98, 99].
89) Hilgetag, G., Teichmann, H.: Angew. Chem. *77*, 1001 (1965); Angew. Chem. Intern.
    Ed. *4*, 914 (1965); Teichmann, H., Jatkowski, M., Hilgetag, G.: Angew. Chem. *79*,
    379 (1967); Angew. Chem. Intern. Ed. *6*, 372 (1967).
90) Märkl, G., Merz, A., Rausch, H.: Tetrahedron Letters *1971*, 2989.
91) Wieland, H.: Liebigs Ann. Chem. *381*, 200 (1911); Neugebauer, F. A., Fischer, P. H. H.:
    Chem. Ber. *98*, 844 (1966); Neugebauer, F. A., Bamberger, S.: Angew. Chem. *83*, 47,
    48 (1971); Angew. Chem. Intern. Ed. *10*, 71 (1971).
92) Kanter, H.: Dipl.-Arbeit, Univ. Marburg/Lahn 1972; Kanter, H., Dimroth, K.: Unpublis-
    hed results.
93) Schmidpeter, A., Ebeling, J.: Angew. Chem. *80*, 197 (1968); Angew. Chem. Intern.
    Ed. *7*, 209 (1968).
94) Hettche, A., Dimroth, K.: Tetrahedron Letters *1972*, 829.
95) Dimroth, K., Hettche, A., Kanter, H., Städe, W.: Tetrahedron Letters *1972*, 835.
96) Dimroth, K., Schäfer, W., Pohl, H. H.: Tetrahedron Letters *1972*, 839.
97) Schäfer, W., Dimroth, K.: Tetrahedron Letters *1972*, 843.
98) Hettche, A., Dimroth, K.: Tetrahedron Letters *1972*, 1045.
99) Dimroth, K., Hettche, A., Städe, W., Steuber, F. W.: Angew. Chem. *81*, 784 (1969);
    Angew. Chem. Intern. Ed. *8*, 776 (1969).
100) Schaffer, O.: Dipl. Arbeit Univ. Marburg 1971; Schaffer, O., Dimroth, K.: Unpublished
    results.
101) Constenla, M., Dimroth, K.: Unpublished results.
102) Schäfer, F. P., Hettche, A., Dimroth, K.: Unpublished results.
103) Guggenheim, E. A.: Trans. Faraday Soc. *47*, 714 (1949); C. P. Smyth, in: A. Weiss-
    berger, Physical methods of organic chemistry, Vol. III, p. 2599. 3rded. New
    York: Interscience Publ. Inc. 1960.
104) Denney, D. D., Relles, H. M.: J. Am. Chem. Soc. *86*, 3897 (1964); Ramirez, R.,
    Lery, S.: J. Am. Chem. Soc. *79*, 67 (1957); look also [83] especially page 70.
105) Daly, J. J., Märkl, G.: Chem. Comm. *1969*, 1057.
106) Thewalt, U.: Angew. Chem. *81*, 783 (1969); Angew. Chem. Intern. Ed. *8*, 769 (1969).
107) Thewalt, U., Bugg, Ch. E., Hettche, A.: Angew. Chem. *82*, 933 (1970); Angew. Chem.
    Intern. Ed. *9*, 898 (1970); Thewalt, U., Bugg, Ch. E.: Acta Cryst. *B 28*, 871 (1972).
108) Schweig, A., Schäfer, W., Dimroth, K.: Angew. Chem. *84*, 636 (1972); Angew. Chem.
    Intern. Ed. *11*, 631 (1972).
109) Dewar, M. J. S., Lucken, E. A., Whitehead, M. A.: J. Chem. Soc. *1960*, 2423.
110) Craig, D. P., Paddock, N. L.: J. Chem. Soc. *1962*, 4118.
111) Mason, S. F.: Nature *205*, 495 (1965).
112) Vilceanu, R., Balint, A., Simon, Z.: Nature *217*, 61 (1968).
113) Gillespie, R. J.: Angew. Chem. *79*, 885 (1967); Angew. Chem. Intern. Ed. *6*, 819 (1967).
114) Perst, H., Dimroth, K.: Tetrahedron Letters *24*, 5385 (1968).
115) Windaus, A., Dimroth, K.: Ber. Deut. Chem. Ges. *70*, 376 (1937); Windaus, A.,
    Dimroth, K., Breywisch, W.: Liebigs Ann. Chem. *543*, 240 (1940); other literature
    cit. in Dauben, W. G., Fonken, G. J.: J. Am. Chem. Soc. *81*, 4060 (1959); Tamelen,
    E., E. van Pappas, S. P., Kirk, K. L.: J. Am. Chem. Soc. *93*, 6092 (1971).
116) Hünig, S.: Angew. Chem. *76*, 400 (1964); Angew. Chem. Intern. Ed. *3*, 548 (1964).
117) Kohler, E. P., Patch, R. H.: J. Am. Chem. Soc. *38*, 1205 (1916); Ziegler, E., Snatzke,
    G.: M. *84*, 610 (1953).
118) Schönherr, H. J., Wanzlick, H. W.: Chem. Ber. *103*, 1037 (1970) and forgoing publi-
    cations; Hocker, J., Merten, R.: Chem. Ber. *105*, 1651 (1972).
119) Wiberg, N.: Angew. Chem. *80*, 809 (1968); Angew. Chem. Intern. Ed. *7*, 766 (1968);
    Hoffmann, R. W.: Angew. Chem. *80*, 823 (1968); Angew. Chem. Intern. Ed. *7*, 754
    (1968).

120) Dimroth, K., Reichardt, Ch., Siepmann, T., Bohlmann, F.: Liebigs Ann. Chem. *661*, 1 (1963); Reichardt, Ch.: Liebigs Ann. Chem. *752*, 64 (1971) (VI. Mitteil.); Reichardt, Ch., Dimroth, K.: Fortschr. chem. Forsch. *11*, 1 (1968/69).

121) Dimroth, K., Umbach, W., Thomas, H.: Chem. Ber. *100*, 132 (1967); Dimroth, K., Laufenberg, J. v.: Chem. Ber. *105*, 1044 (1972).

122) Pitz, H.: Dipl.-Arbeit, Univ. Marburg 1972.

123) Pohl, H., Dimroth, K.: Unpublished results.

Addition in proof (March 23, 1973):

124) Märkl, G., Heier, K. H.: 1.1-Dibenzyl-2-phenyl-phosphanaphthaline. Angew. Chem. *84*, 1066 (1972); Angew. Chem. Intern. Ed. *11*, 1016 (1972),

125) Märkl, G., Heier, K. H.: 2-Phenyl-1-phospha-naphthaline. Angew. Chem. *84*, 1067 (1972); Angew. Chem. Intern. Ed. *11*, 1017 (1972).

126) Märkl, G., Matthes, D.: 2.6-Diphenyl-1-aza-4-phosphabenzene. Angew. Chem. *84*, 1069 (1972); Angew. Chem. Intern. Ed. *11*, 1019 (1972)

127) Märkl, G., Fischer, D. E.: Methylen-phosphacyclohexadien-phosphabenzol-Umlagerung; Zum Mechanismus. Tetrahedron Letters *1973*, 223.

128) Kuczkowski, R. L., Ashe, A. J., III: The microwave spectrum, dipole moment and low frequency states for phosphabenzene. J. Mol. Spectr. *42*, 457 (1972).

129) Kanter, H., Dimroth, K.: 1.1-Dihalogen-λ^5-phosphorins. Angew. Chem. *84*, 1145 (1972); Angew. Chem. Intern. Ed. *11*, 1090 (1972),

130) Schaffer, O., Dimroth, K.: Reactions of λ^3-phosphorins with arenediazoniumsalts to λ^5-phosphorins. Angew. Chem. *84*, 1146 (1972); Angew. Chem. Intern. Ed. *11*, 1091 (1972).

131) Batich, C., Heilbronner, E., Hornung, V., Ashe, A. J., III., Clark, D. T., Cobley, U. T., Kilcast, D., Scanlan, I.: Photoelectron spectra of phosphabenzene, arsabenzene and stibabenzene. J. Am. Chem. Soc. *95*, 928 (1973).

132) Hettche, A., Dimroth, K.: Autoxidation und Wasserstoffperoxid-Oxidation von 2.4.6-Triphenyl-λ^3-phosphorin. Chem. Ber. *106*, 1001 (1973).

133) Fraser, M., Holah, D. G., Hughes, A. N., Hui, B. C.: Donor properties of 2-coordinate phosphorus in the phosphorin system; Metal complexes. J. Heterocyclic Chem. *9*, 1457 (1972).

Received August 23, 1972

Springer-Verlag
Berlin Heidelberg New York
München London Paris Sydney Tokyo Wien

C. Fest
K.-J. Schmidt

The Chemistry of Organophosphorus Pesticides

Reactivity – Synthesis – Mode of Action – Toxicology

With 46 figures.
X, 339 pp. 1973.
Cloth DM 88,–;
US $ 32.60

This book about the chemistry and mode of action of organophosphorus insectics lays particular stress upon the way in which activity is related to structure. It has three main themes:

1) How the basic types of pesticides are produced and the chemical know-how needed to do this

2) How definite compounds are prepared; their chemical structure and applications

3) Mechanisms of action

Both authors are actively engaged in research on chemical pesticides; their approach thus represents the chemist's point of view and is consequently likely to interest chemists working in other fields. Moreover, since plant protection is nothing if not interdisciplinary, this book can offer a useful basis for biologists, entomologists, mycologists, botanists, farmers and adminstrators concerned with agricultural research and environmental studies.

Beyond this, the authors had the definite intention of writing a book which, by pointing out how complex are the problems associated with pest control, will, they hope, inject a certain objectivity into the public debate on these matters.

Prices are subject to change without notice

Contents: General Section. - Chemical Section. - Biochemistry.

J. Falbe

CarbonMonoxide in Organic Synthesis

Translated by C. R. Adams
With 21 figures. IX, 219 pages. 1970.
Cloth DM 58,–; US $ 21.50

The importance of carbon monoxide chemistry has increased rapidly in the last few years, both in scientific research and chemical processing. This necessitated the revision of the book, Synthesen mit Kohlenmonoxyd, published in German in 1967 as Vol. 10 in the series „Organische Chemie in Einzeldarstellungen". This covered Roelen's discovery of hydroformylation or oxo reaction, Reppe's carbonylation process, Koch's carboxylic acid synthesis and ring closure with carbon monoxide. The new edition includes latest research findings in these fields and, in particular, more space is given to the discussion of reaction mechanisms. The latest developments in industry are also mentioned and there have been numerons additions to the list of references.

This book will be an important reference source, both for the established expert and for those who wish to enter the fields of petrochemicals, organic chemicals and chemical engineering, particularly as results tend to be published in patents and thus remain outside the ken of a wide range of readers.

Contents: The Hydroformylation Reaction (Oxo Reaction/Roelen Reaction). - Metal Carbonyl Catalyzed Carbonylation (Reppe Reactions). - Carbonylation with Acid Catalysts (Koch Reaction). - Ring Closures with Carbon Monoxide. - Laboratory Preparations with Carbon Monoxide.

Prices are subject
to change without
notice

**Springer-Verlag
Berlin Heidelberg New York**
München London Paris Sydney Tokyo Wien

In kritischen Übersichten werden in dieser Reihe Stand und Entwicklung aktueller chemischer Forschungsgebiete beschrieben. Sie wendet sich an alle Chemiker in Forschung und Industrie, die am Fortschritt ihrer Wissenschaft teilhaben wollen.

In der Regel werden nur Beiträge veröffentlicht, die ausdrücklich angefordert worden sind. Schriftleitung und Herausgeber sind aber für ergänzende Anregungen und Hinweise jederzeit dankbar. Manuskripte können in den „Fortschritten der chemischen Forschung" in Deutsch oder Englisch veröffentlicht werden.

Jeder Band der Reihe ist einzeln käuflich.

This series presents critical reviews of the present position and future trends in modern chemical research. It is addressed to all research and industrial chemists who wish to keep abreast of advances in their subject.

As a rule, contributions are specially commissioned. The editors and publishers will, however, always be pleased to receive suggestions and supplementary information. Papers are accepted for "Topics in Current Chemistry" in either German or English.

Any volume of the series may be purchased separately.